The Future of
Electrical Energy

The Future of
Electrical Energy ─────────
A Regional Perspective of ────
An Industry in Transition ────

Edited by
Sidney Saltzman
and
Richard E. Schuler

New York
Westport, Connecticut
London

Library of Congress Cataloging-in-Publication Data

The Future of electrical energy.

Includes index.
1. Electric power — Congresses. I. Saltzman, Sidney.
II. Schuler, Richard.
TK5.F88 1986   333.79′32   86-533
ISBN 0-275-92158-1 (alk. paper)

Library of Congress Catalog Card Number: 86-533
ISBN: 0-275-92158-1

First published in 1986

Praeger Publishers, 521 Fifth Avenue, New York, NY 10175
A division of Greenwood Press, Inc.

Printed in the United States of America

∞
The paper used in this book complies with the Permanent
Paper Standard issued by the National Information Standards
Organization (Z39.48-1984).

10 9 8 7 6 5 4 3 2 1

Contents

List of Tables ————————————

List of Figures

Foreword

It seems to me that we are in the midst of a very serious change in our society, one that bears upon each of the issues discussed at the conference that is the subject of this book.

Simply put, society's view of what government is has changed. For many years the government was a straightforward representative democracy. We elected people from our local towns, sent them to Washington or to Albany and asked them to represent us. We let them go ahead and do it until they came back for a new grant of authority from the citizens. That grant was either given or denied without a whole lot of hoopla.

But we seem to be moving increasingly toward direct democracy in the classical sense. Less and less are we trusting anybody in the government to make our decisions for us. A signpost of this change is the proliferation of referenda across the country on every conceivable issue. In part the public is withdrawing the grant of power. People want to vote practically every day on whether or not their elected representatives are doing the right thing. And in part, the change has come about because the politicians find it convenient to duck certain issues. They cannot decide or do not want to decide what to do about difficult issues, such as abortion or re-allocation of hydropower, so they let the public vote on it in a referendum. That way they avoid, at least to some extent, the decision making and the anger of both sides.

On the other hand, it is often more dangerous when the legislators do act. Thomas Jefferson did not trust politicians who belonged to a special class. He hoped we would end up with ordinary citizens in the government. But today we no longer have people who work part time and then go to Congress; they are in Congress the whole year. They do not come back and farm or trade as they did in the old days. Instead, they come back and tell you what a great job they have been doing. This leads to what I call the "legislative imperative." "We cannot just sit there," they say, "we have to pass something."

The legislators cannot go home and say that all the bills were useless, and therefore they did not pass any of them. This lacks spunk and sizzle when you are trying to explain to the New York *Times* what you did for the last year for your $75,000 salary, plus another roughly quarter of a million dollars for staff. So members of Congress have to come back with a whole list of things they passed.

As a result we now have more and more rules and regulations, and the public is rebelling and taking the Tenth Amendment seriously. They want the federal government to do only those things that have been directly delegated to it. The rest should be left to the states or to the people.

Such is the spirit of the American public today. More and more we are withdrawing from the government the inherent trust we used to have, trust that is necessary to a representative democracy. We are making sure that every decision is checked and balanced and checked and balanced again. This has led to a kind of paralysis. We cannot get 220 million people in a room and have a chat, which was possible in the classical democracies of Greece and Italy.

I am concerned about this replacement of representative democracy by this as yet unfocused system that resembles a direct democracy. I fear that our society does not yet have the tools to arrive at the necessary and correct judgments nor the ability to make them work for all of us.

What do we in the energy business do? On the one hand everything we want to do is unwelcome somewhere. Nobody wants a power line or a power plant. Nobody wants to change from oil to coal, which might be dirtier. People do not want these things built in their backyard. But they absolutely expect, as a matter of simple assumption, that the electricity will work when they turn on the switch. Nobody in the political world seems to pay any attention to the necessity of doing certain things to ensure that when the switch is thrown the juice is there.

How do we organize ourselves in this increasingly participatory democracy? How do we organize ourselves to arrive at decisions on complicated issues which are enmeshed in technocratic jargon that is hard even for professionals in other fields to understand, much less the ordinary citizen? How do we get a decision that makes some sense? How do we get people in the political arena to understand that they must have some mechanism for arriving at a decision? And that once they have reached a decision they must allow it to be implemented?

How do we organize ourselves to build a coal plant that benefits nine million people and discomforts no more than 200,000, one that will replace Arab oil, save money, and eventually pay for itself? We have failed in New York to discover a mechanism to allow this to happen. I hope we do not fail to find the mechanisms that will allow Canadian hydropower to replace oil—one example of the items on our agenda. We undertook the conference to see if we could find a better way. I trust that the views and insights of the distinguished participants collected in this volume will point us in the right direction. New York's electric future depends on it.

The conference and this volume are the result of a productive collaboration between the electric utility industry and the academic world. The partners were uniquely suited to the task of planning and presenting a major program focused on the diverse and complex issues that will shape the state's electric future. The New York Power Authority is the state's largest supplier of electricity and is at the forefront of various efforts to meet our current and future energy challenges. Cornell University's distinguished faculty includes several members with a special interest and expertise in the energy field, including a former member and a former chairman of the New York State Public Service Commission.

Cornell and the New York Power Authority, as cosponsors of the conference assembled an impressive array of knowledgeable speakers and an equally formidable audience. In this endeavor, we had the invaluable assistance of the New York State Energy Office and the New York City Bar Association, in whose fine facility the conference was held.

The effort bore fruit as 300 economists, energy planners, government and electric utility officials, academicians, and journalists participated in the two-day program. The exchange was informative, balanced, and provocative. That spirit has been captured in the pages that follow.

John S. Dyson
Former Chairman
New York Power Authority
Millbrook, New York

Preface

Many of the problems related to the generation, transmission, and use of electrical energy are formulated and dealt with at the state or regional level. Furthermore, the basic nature of these problems is, to a considerable extent, common to a wide variety of states and regions of the United States. For example, virtually all states and utilities have similar forecasting problems; environmental and health effects are common throughout various multistate regions in the United States; and the technological problems of generating and transmitting electricity are more similar than different for many utilities. For these reasons, we believe that analysis of the future of electrical energy for a major industrial state bears on other states, regions, and utilities in the United States.

This volume contains the edited papers presented at a conference entitled "New York's Electrical Future" that was held on November 8 and 9, 1984, in New York City. The purpose of this conference, which was jointly sponsored by Cornell University and the New York Power Authority, was to examine the future of electrical energy in New York State from the following perspectives:

- the long-range demand for electricity and the relationships among electricity use, prices, and the economy;
- current sources of supply of electricity and factors influencing their use;
- alternative technologies for future sources of supply of electricity;
- environmental and health effects of alternative methods for the generation and transmission of electricity;
- public perceptions and attitudes toward electricity generation and use;
- financial aspects related to the supply of electrical generating facilities;
- regulatory and institutional issues related to the supply, distribution, and pricing of electricity.

In order to focus the attention of the conferees on each of these topics and provide a comprehensive overview of the diverse but related issues, the sessions were scheduled to run consecutively during the two days of meetings. Because there were no parallel sessions, the conference provided many opportunities for specialists and those with more general interests to interact with each other.

Attendance at the conference was by invitation. The more than 300 conferees were drawn from among thought leaders in the various professions, academia, labor unions, government agencies, and the media, and from among consumer groups, all of whom have had more than a casual interest in at least some aspect of the electric utility industry.

In most cases, drafts of the speakers' papers were available at the conference. All of the proceedings were recorded and transcribed, including comments from the floor and also questions and answers. Each speaker then had the opportunity to revise his or her paper based on the written transcription and any related discussion. Thus, these edited papers contain most of the important ideas, opinions, and summaries of discussions presented at the conference.

The main purpose of the conference was to raise important issues about the future of electricity in New York State and to enlighten the debates on these issues. It was too much to expect that solutions for even some of the underlying problems would be agreed upon in a two-day conference that was designed to be comprehenseive rather than sharply focused. The editors as well as the staff members from the New York Power Authority who played important roles in the organization of this conference trust that the publication of these proceedings will "enlighten the debates on these issues."

There are a number of people at the New York Power Authority and at Cornell University who played important roles in planning and executing the conference and in bringing this volume to fruition. In particular, we would like to recognize the contributions of and thank the following people at the New York Power Authority: John Dyson, chairman of the Power Authority at the time the conference was held, and James Cunningham, senior vice president for public affairs who originally sensed the need for and promoted the conference. Terrence Curley, Anne Strauss, and Anthony Schillizzi of the Power Authority also had important roles in organizing and running the conference and we thank them for their special expertise and assistance. We also are indebted to the staff of the Cornell Institute for Social and Economic Research for their administrative support. We especially would like to thank Careen Morse for the very cheerful, pleasant, and highly competent way she prepared this manuscript and performed various administrative tasks connected with this project. Two further points should be noted for the record. First, the opinions expressed in the manuscript are those of the authors of the respective papers and not those of the New York Power Authority or Cornell University. Second, we are responsible for opinions expressed in the introduction and epilogue and in the connecting sections that introduce each major group of chapters. We, of course, are also responsible for any errors that may remain in the manuscript.

The Future of
Electrical Energy

Introduction

Sidney Saltzman
Richard E. Schuler

The electric power industry is a cornerstone of modern America; yet in recent years it has been buffeted by economic and international events, attacked by politicians, and reviled by some members of the public to an extent not experienced since the 1930s. Nevertheless, the electric generators continue to hum, the demands of new customers are being met, and the industry has moderated the level of pollution it emits. Furthermore, the recent continuous waves of rate increases, induced in part by the 1974 and 1978 oil embargoes, have also receded for many utilities over the previous two years. And although the demand for electricity continues to grow, the rate of increase has decreased significantly in most regions of the country, reducing the need for the utilities to mount massive construction programs to meet future demand. Besides, most utilities acknowledge that they have no further plans to initiate new nuclear construction programs in the foreseeable future, thereby eliminating a primary source of the public's anxiety and irritation with the industry. Why then is there any need to provide a postmortem, let alone plan and rethink the future of the electric industry?

There are three simple answers to this question. First, a quick review of previous forecasts for the electric power industry, based upon simple extrapolations from recent experience, reminds us that those projections often have been wrong. Second, this industry is absolutely essential to the well-being of modern society. In one sense, the industry performed too well up until the last decade, leading most Americans to take it for granted, but as recent events have demonstrated, any shortcomings in this industry can lead to substantial adverse consequences for everyone. Third, the electric industry is the most capital intensive of any in the U.S. economy, and because of its scale and the broad public and environmental consequences of erecting most new facilities, the planning and construction period now requires a 10–15 year lead time. Taken together, what these three factors imply for the future of electricity supply is summarized quite succinctly by James Schlesinger: sequential periods of complacency followed by panic. The fact that we may be entering into a period of relative calm, though a welcome respite from the previous decade, also provides us with an invaluable opportunity to rethink the future and perhaps to avoid or reduce the magnitude of the next panic. That is the object of this volume.

Although these issues are of nationwide importance, exploring them within a particular regional context allows broad concepts to be examined and tested in light of specific circumstances. This volume regards the particular experiences of and projections for electricity supply and demand, and of regulatory, financial, and environmental constraints in New York State as case illustrations of larger nationwide problems. Furthermore, many decisions about the generation, transmission, and use of electrical energy are made on a regional, state, or substate rather than national level. Most electric utilities service regions smaller than entire states and, typically, are regulated by one or more state agencies operating under the authority of the state government. Although the federal government intervenes on issues that transcend state or regional interests and/or boundaries (e.g., for reasons of regional supply coordination, wholesale power purchases, environmental, health, or safety), many of these issues are resolved at the state or regional levels. Thus the government's concern includes the demand, supply, and price of electricity, its environmental and health effects, and the financial and regulatory aspects of the utilities.

The first three sections of this volume focus on the interrelationship between electricity supply, its reliability and price, and the economic health of specific regions. The final four sections discuss constraints that may influence the availability and direction of alternative sources of electricity.

Section I provides a global overview of energy problems and the role of electricity—its history and economic impact. Richard Schuler opens this section with a brief history of the electric utility industry and its evolution in the United States and in New York State. Lawrence Klein provides a restrospective of the energy crisis of 1974 and discusses what can be learned from that experience. James Schlesinger discusses the long-run security of electricity supplies and how that may be influenced by the actions of other countries. He also analyzes the future role of oil imports from the Middle East and hydro-power from Canada, together with the likelihood of a future crisis in the electricity supply.

The chapters in the second section develop demographic and economic forecasts for New York State and integrate these with projections of electricity demand. Nancy Meiners presents baseline economic and demographic forecasts for the region, which form the basis for the electricity forecasts for New York State. Nicholas Johnson and John Adams who also emphasize the differing patterns of growth among the various utilities within the state present electricity demand forecasts from the New York Power Pool. State Energy Office forecasts for energy, including electricity, are presented by Charles Guinn. Timothy Mount and William Deehan compare a range of past electricity demand forecasts, and they outline how alternative environmental regulations dealing with acid rain and how alternative legal and accounting treatment of the costs of completing two nuclear plants in New York might influence electricity prices and therefore future load growth. This section concludes with an assessment of world oil markets and the energy needs of particular industries by Douglas Bohi and Joel Darmstadter. These authors relate oil prices to electricity prices and show which industries are consequently most sensitive to changing energy and electricity prices.

The third section examines the ability of existing and projected electricity supplies to meet these forecast demands. William Davis gives the State Energy Office's overview for New York, and Lester Stuzin reviews potential operating problems from the Public Service Commission's perspective. Robert Hiney evaluates the potential for additional imports of Canadian hydro-power; Hiney is followed by Robert Percival, who assesses how much of New York's demand may be satisfied by additional conservation measures and by electricity from small, nonrenewable sources and cogeneration.

The final two papers in this section extend the discussion of alternative supply sources by exploring the (potential) role of emerging technologies. Alvin Weinberg examines the likelihood that alternative forms of nuclear power will play a role in future electricity supplies, and he summarizes the debate between proponents of large- and small-scale solutions. Frank Huband evaluates the potential applications of emerging alternative electricity supply and conservation technologies.

In the fourth section, existing and potential future environmental constraints are reviewed. Senator Robert Stafford offers a federal policy perspective on environmental regulation. The particular problems of fossil fuel-fired plant emissions for New York State, including acid rain, are reviewed by Harry Hovey. William Stasiuk discusses the public health implications of alternative electricity supply sources. Finally, Alan Crane, in a review of the particular safety problems of nuclear power, enumerates and analyzes the environmental, health, and safety problems that may arise from each of the alternative means of electricity generation and transmission.

In the fifth section, Donald DeLuca and Steven Barnett provide two independent assessments of public perceptions and attitudes toward various sources of electricity and their supplying institutions. Howard Axelrod presents a methodology for predicting future regulatory crises.

Financial needs and constraints are the topic of the sixth section, which begins with a projection of the financial strength of New York's investor-owned utilities by Leonard Hyman, Doris Kelley, and Richard Toole. Philip Kron provides the perspective of commercial bankers who are relied upon to provide funds during construction. Ronald Forbes examines the financial strength and cost of capital for public power suppliers, and the New York Power Authority in particular.

In the final and concluding section, regulatory and institutional constraints are reviewed and possible changes are proposed. Richard Schuler analyzes the interrelationship between economic and environmental regulations, and develops a theoretical basis for change. Charles Stalon examines the regulatory decision-making process from the federal perspective and suggests modifications in procedures. And Alfred Kahn offers his own assessment of needed regulatory changes and critiques proposals in preceding chapters. In the final chapter the volume editors provide a summary and analysis of selected issues of major importance to the future of the electric utility industry.

PART I

ELECTRICAL ENERGY: INTRODUCTION AND OVERVIEW————————————————

The stage is set by the authors of this section for the subsequent discussions of critical issues that face the electric power industry. One perspective for understanding is knowing how the industry, its supply technologies, the use of its product, its institutions and regulatory philosophy developed over time. Richard Schuler provides a brief review of these topics in Chapter 1. It is interesting to note that although the problems have changed, the utility industry has been the focus of controversy for much of its first 100 years.

Lawrence Klein focuses on more recent historical developments: the energy crises of the 1970s, which changed the patterns of relationships among people, institutions, and nations, much as did the development of electric power during the previous century. Instead of tracing the chronology of the major events that constituted the crises, he interprets their more important aspects. For example: What was the nature and meaning of the crises? Were they "real" or merely the result of people's perceptions? What were their major impacts in the United States and internationally?

Klein's review of the U.S. response to the energy crises and how changes in the relative prices of energy affected the economic activities of various constituencies within the United States traces the economic outcomes of the responses to the crises up to the present time. Finally, Klein shifts his attention to New York State and considers the impacts of the energy crises on the economy of the state and the conditions that are likely to influence New York's future economic growth. Thus, Klein and Schuler set the stage for the more detailed analysis of the energy–economy interactions that are presented in the subsequent sections.

James Schlesinger's point of departure for his analysis of the future security of energy supplies is also the energy crisis of the 1970s, because it was the security issue that those events brought to the fore. Although both Schlesinger and Klein are concerned primarily with the economic aspects of energy, Schlesinger addresses the long-term future of oil and electricity supplies.

Many of the points introduced by Schuler, Klein, and Schlesinger in their three papers are addressed in considerably more detail in the remaining sections of this volume. Energy–electricity–economy interactions and projections of future electricity demand are examined in detail by the authors in Part II, and a variety of supply issues are elaborated on in Part III. Thus Part I establishes the setting and an introduction to major issues of regional electric power supply, demand, and security, which will then be developed in detail throughout the remainder of this volume.

1

Electricity in New York State: The First One Hundred Years as Prelude to the Future ──────────

Richard E. Schuler

Few would deny that the electric utility industry is currently at a crossroads; indeed, some individuals would contend it has reached its nadir and is currently a seriously troubled institution. As evidence can be cited the dramatic post-World War II reduction in demand growth, from 7–8 percent annually through 1970 to 1–2 percent over the past decade. In addition the industry has been caught in the energy maelstrom following the 1974 oil embargo, as a result of which nominal prices, which regularly fell in previous decades, have now been replaced by regular annual requests for rate increases (a recent Consolidated Edison [ConEd] action being a notable exception). Finally, the industry's technology of the future, nuclear fission power, which was purported to be able to provide an unbounded supply of electricity at continually lower prices, is currently perceived by many as a failure. Under these circumstances, why should any imaginative assessment of the future pause to consider the role of electricity?

The overriding reason for being concerned with New York's electric future is the simple, irrefutable fact that modern society cannot exist without electricity. Whether or not this dependence will continue to grow is an issue that the chapters in this volume will address. However, merely sustaining the existing level of supply at moderate cost is no simple task in the face of the ravages caused by inflation over the past decade. Furthermore, for the last several years we have observed a modest but persistent annual growth of 1–2 percent in New York State's use of electric-

I am deeply indebted to Lewis Tatem who assisted me in identifying many of the key events in the following chronology and in confirming their sources. Decisions regarding which events to emphasize in this brief summary are entirely my own, and therefore the many salient oversights are also my responsibility.

ity. When extrapolated beyond the current glut of excess capacity, this increase translates into a need for 500 megawatts (MW) annually of new generating capacity to serve New York State by the middle of the next decade. Recent history emphasizes the foolhardiness of simply extrapolating past usage patterns in order to forecast future demands; nevertheless, the mere fact that the current planning and construction horizon needed to bring a new central station facility on-line is equal to or exceeds that ten-year demand projection makes it equally foolhardy not to examine the reliability of and consider the possible implications of those extrapolations. Many of the chapters that follow attempt to provide an underlying understanding of the reasons for this recent moderate growth in demand.

Although electricity generation may not be generally viewed as a primary growth industry, many of the functions it serves, like telecommunications advances and enhanced, individualized computation power, both within service industries and for robotized manufacturing systems, are definitely seen as the wave of the future. While the electrical needs associated with making these substitutions for human brainpower are modest, it is equally undeniable that some electricity input is absolutely essential. And, to the extent that they do substitute for human brainpower, the electric power required for their operation must be at least as reliable as our own powers of concentration. Indeed, one of the major impetuses over the past decade for the substantially improved reliability of electric service was the need of mainframe computer systems for steady, uninterrupted electricity supply.

In order to gain a broader perspective for understanding the future of electrical utilities, this chapter provides a brief retrospective of the electric power industry and its relationship with other technologies. While the past may not be prologue to the future, it is necessary to understand why and how we arrived at the current situation in order to determine whether or how we might want to synthesize a different institutional and regulatory framework for the future. Thus I will also review highlights of the technological and institutional evolution of the supply and use of electricity in New York State and the United States. The several conceptual threads that have persisted throughout this history set the stage for outlining the key topics addressed in this volume.

HISTORICAL TECHNOLOGICAL AND INSTITUTIONAL EVOLUTION OF ELECTRIC POWER SUPPLY IN NEW YORK STATE

Evolving Technology

Charles Brush may not be a household name; nevertheless, he is responsible for the first commercial electric power supply system in New

York State, which became operational on a contract basis in 1881.[1] Unfortunately, this was not the first commercially operated venture in the United States: that honor goes to San Francisco, where a Brush system began commercial sales in 1880. Ironically, Brush's New York City system was actually installed in 1880, but the city fathers were reluctant to accept this venture without a one-year trial. Brush's New York system consisted of twenty-two arc streetlights stretched along Broadway between 14th and 34th streets. The system was powered by direct current (DC) and was laid out in a series circuit so that if one light failed, all of Broadway was thrown into darkness.

Meanwhile, Thomas Edison, who had developed the incandescent lamp in 1879, was working feverishly on a direct-drive DC generation system to power his lighting system. Brush's and all earlier generation systems were belt-driven and therefore substantially less efficient than Edison's system. Furthermore, Edison proposed a parallel circuit system for distributing electricity, which was installed together with his famous Pearl Street generating station in 1882 to serve the Wall Street area of New York City. Both because of the greater reliability of his distribution system (when one light went out, the others remained lit) and its improved efficiency, Edison is usually thought of as the initiator of commercial electric power supply in the U.S.

Following the initial success of Edison's system, the great AC-DC debate erupted and raged over the next twenty years between two titans of the electric supply industry, Thomas Edison and George Westinghouse. Advantages claimed for the alternating current (AC) system were more efficient, trouble-free generation and, more significantly, the ability to alter the voltage of the supply through use of a transformer, thereby permitting transmission of power over much longer distances with fewer losses. The first commercial AC system was installed in Buffalo in 1886, and it transmitted power at 1,000 volts as compared with Edison's 110 volt system.[2] Nevertheless, the combination of Edison's strong advocacy, plus the ease with which nonscientific users could conceptualize his system, maintained substantial support for Edison's DC system over many years. In fact these early pioneers in the industry engaged in a rough-and-tumble, no-holds-barred debate that makes today's colloquy between nuclear power, coal-fired plants, and soft energy and conservation advocates look like a gentlemanly joust. Edison's supporters alleged the inherent unsafety of the AC system, alluding to the possible failure of transformers, and that such a failure would cause high-voltage electricity to be delivered directly into individual homes. This debate gravitated to its lowest level at the end of the nineteenth century when, stirred by a particularly grisly execution in Utah, most states in the country were searching for more "humane" forms of capital punishment. Electrocution became a prime candidate, and in 1889 New York State purchased two AC generators to power the state's first electric chair. Edison's advocates immediately proposed that this method of capital punishment be named "Westinghousing" in honor of that gentleman of science.[3]

Other early technical factors to be decided were whether AC current should be 25 or 60 cycles (in Europe it is currently 50 cycles) and whether polyphase service should be two or three phase. Such fundamental decisions make today's discussions over standardization of outlets for telecommunications equipment and the telephone jacks seem trivial. With the rapid adoption of AC systems, the interconnection of neighboring generating companies in order to enhance system reliability at a cost lower than that required to duplicate facilities was the next step in the evolution of electric systems. In 1896, the Niagara Falls Power Company completed the first interconnection in the U.S. by constructing a 22 kilovolt (kV) line to span the 20 miles between its generators and the electrified Buffalo Railway Company. This interconnection was also the first to add a degree of supply diversity, since the Niagara Falls generation was hydroelectric; the Buffalo Railway Company, on the other hand, used steam-powered generation. The normal economical flow was from Niagara Falls to Buffalo; however, when in 1902 the Niagara Falls generation was taken out of service for routine maintenance, the practical feasibility of reverse-flow was demonstrated.[4] Other interconnections quickly followed. In 1923, the Central Hudson Electric Company and the Adirondack Power Company completed a 50 mile transmission link that was subsequently extended to New England,[5] and in 1925 the Niagara Power Corporation was the first to build a transmission line rated at over 100 kV.

The major contemporaneous advances in the generation of electricity were in the steam-powered machines used to turn the generator's shaft. Early designs relied on previously available steam engines. As an example, in 1900 New York City's Grand Central Station had two engines, each capable of producing 35,000 kW; however, each stood 65 feet tall, weighed over 500 tons (450,000 kilograms), and occupied 1,000 square feet of floor area.[6] The subsequent development of steam turbines meant that the same power could be produced utilizing one-fifth the floor space, and the much greater rotational speed was highly compatible with high-voltage AC generation. Thus by 1915, 35,000 kW steam turbine generators were also available, and the maximum output of turbine-driven generators escalated rapidly to 60,000 kW by 1920, 90,000 kW by 1925, 165,000 kW by 1936,[7] and on to the 1 million kW units put into operation in the post-World War II era.

During this period the average cost of generating electricity declined continuously because of technological and operational improvements that resulted in substantial scale economies. However, concomitant advances in transmission technology, which through ever higher voltages allowed greater quantities of electricity to be transmitted longer distances, were necessary for the economic utilization of those scale economies in generation. As an example, given a fixed demand for electricity by each consumer, larger more efficient generating units would have to transmit farther to serve more customers, if their total capacities were to be fully utilized. But this required improvements in the ability to transmit power at reasonable costs. Thus both technologies advanced in tandem, with improvements in one paving the way for subsequent advances in the other.

Evolution in the Demand for Electricity

From its inception, few users were able to foresee completely the multiplicity of uses and the value of electricity for industry, business, and the home. The original dominant advantage of this source of energy was the illumination it provided, and today electricity still occupies a natural monopoly status for this function because of the low cost with which it can provide flexible high levels of illumination. Electricity's next major inroad was as a substitute for water or direct-drive steam systems as a source of rotational power in industry. Because of the massive bulk of these previously nonelectric, direct-drive systems, all machinery had to be aligned in straight rows; a common shaft or attached belts turned the production machinery. While the first electric motors were merely substituted as prime movers for those direct-drive power sources, it was quickly recognized that many small motors, one attached to each piece of machinery, would permit a much more flexible layout of work space. As pointed out by Warren Devine,[8] this is the first instance in which a direct extrapolation of historical use patterns woefully underestimated the burgeoning demand for electricity. True, less installed horsepower was required when one large central motor was connected to all of the machines in a factory through a complex mechanical-drive system. However, this capital cost advantage proved to be far outweighed by reductions in labor, material, and building space costs if individual motors and the flexibility of wires used to rearrange equipment with an eye to efficient labor and material flows. Ironically, today the same process seems to be repeating itself as individualized microcomputers are replacing multiple-access terminals to a single centralized main-frame computer.

In fact, three entirely new industries have emerged over the past 100 years solely because of the invention and economical supply of electricity. The first was the introduction of the telegraph and telephone as sources of communication. Next, electrically-driven chemical processes developed solely as the result of the availability of large supplies of electricity.[9] Not suprisingly, early facilities were located near Niagara Falls in New York State. While not nearly as power-intensive, the recent information and computation explosion is equally dependent upon a stable, *reliable* source of electricity supply.

The third widespread adaptation of electricity, for process heating in manufacturing and space-heating applications, is merely a substitute for other primary sources of energy; however, one of the advantages again credited to electricity is its ability to apply the heat precisely where it is needed, in controlled quantities. And air conditioning systems would still be extremely primative without electricity—driven fans to circulate the air, if not to power the chillers.

Evolution of Supply Institutions and Their Regulation

From the initiation of Charles Brush's earliest system, electricity has been supplied through a vertically integrated entity, where the distributor of power has also been its manufacturer. And, during the first twenty-five years of the industry's development, minimal public regulation was introduced since the service was considered a luxury and many competitive companies existed. Nevertheless, several municipalities also began to provide electric service, the earliest being the Dunkirk, New York system in 1888. However, by 1905 the use of electricity in New York State had become sufficiently widespread and the monopolistic abuses by providers of the service sufficiently evident that Governor Charles Hughes proposed the formation of the Public Service Commission.[10] This law was enacted by the legislature, which formed two distinct commissions: one serving New York City and the other for the rest of the state, each of which was empowered to regulate gas, electricity, and railway service within its respective jurisdiction. While the initial thrust of regulation was merely to fix reasonable rates, other operating practices were eventually covered as well. Thus the power of regulators was rapidly expanded to include many functions that persist today: examining books; prescribing uniform accounting and financial reporting systems; requiring the provision of adequate and impartial service; establishing rates; and (in my view the most important function) approving the issuance of all financial securities. In 1921, these two geographically separate bodies were combined to form one statewide Public Service Commission (PSC).

At the same time, with substantial innovations in high-voltage transmission facilities permitting the interconnection of geographically far-flung systems, major forces to consolidate small independent supply systems came into play. Between 1921 and 1930, the 373 independent electric systems in New York State were consolidated into 152 firms. By 1946, there were still 8 separate utilities serving New York City alone, although 4 were controlled by ConEd, and 12 other companies served the remainder of the state. In 1951, 6 of those remaining upstate companies merged to form the present Niagara Mohawk Corporation, and all 8 New York City companies combined into the existing ConEd Company. Thus the present landscape of private utility companies is only thirty-three years old.

Meanwhile, beginning during the Depression, substantial pressure emerged to extend service to outlying rural areas by forming public power corporations. This was the major purpose of the Rural Electrification Act of 1933. Also, in 1931 the enabling legislation was passed in New York State to form the Power Authority of the State of New York (now the Power Authority, or NYPA), which was originally intended to develop hydroelectric generation resources along the St. Lawrence Seaway and the Niagara River. An original and continuing provision was that all of these power projects were to be self-sustaining and not to tap the state's general purpose budget revenue sources. Because of a variety of legal battles with private utilities along the Niagara River and with Canada along the St. Lawrence Seaway, major construction did not begin on any of the NYPA's hydroelectric projects until 1954. In fact, the St. Lawrence project required the ratification of a new treaty with Canada by the U.S. Congress, which faced substantial opposition by coal and railroad interests.[11] However, after the privately owned Schoellkopf station collapsed in 1956, the NYPA's construction program moved into high gear.

In recent years, New York State has continued to provide many firsts in the supply and regulation of electricity. As an example, ConEd's cash-flow problems of 1975, which were coincident with New York City's economic difficulties and an ambitious construction project including nuclear plants at Indian Point, threatened the utility's financial viability. Thus the NYPA's rescue of ConEd, effected by purchases of several of its plants, may indeed have been an early precursor of recent quandaries associated with many nuclear facilities around the country.

On the positive side, in 1972 Articles 7 and 8, appended to the New York State Public Service Law, established siting boards in an attempt to integrate environmental and economic utility regulation through a one-step approval process for transmission lines and new generating stations. Previously up to seventeen state and fourteen federal permits were required in order to construct a nuclear facility.[12] And in 1976, thanks to the heroic efforts of Alfred E. Kahn, and using the commission's legislative powers after an exhaustive set of generic hearings, the PSC began to implement marginal cost pricing for the state's utilities. In retrospect, this landmark pointed to the end of an era, because while marginal cost pricing had always made theoretical sense to economists, it became an absolute necessity as electric utilities were making the transition from decreasing to constant—or increasing—cost industries. Finally, 1977 saw the formation of the State Energy Office, which was an attempt to integrate all aspects of energy supply planning of both regulated and competitive industries.

Aside from the Rural Electrification Administration financed cooperatives (REA), federal involvement in the regulation of power planning had its roots in the amended Water Power Act of 1920.[13] While the initial regulation was related to conservation efforts, the powers of the Federal Power Commission, now the Federal Energy Regulatory Commission (FERC), were gradually extended. Today this organization regulates and approves all wholesale transactions among both inter- and intra-state systems in the country. The federal government is also directly involved in the generation and transmission of electricity through the Tennessee Valley Authority and the Bonneville Power Administration (the most notable examples), but neither of these agencies markets power in New York. Furthermore, both of these federal agencies were initially charged with developing hydroelectric sites; however, they eventually also constructed steam-powered facilities to meet the expanding demand of their customers. In this respect, the evolution of the role of the NYPA in New York has been similar. Since World War II, the federal government has set the stage for the commercial development of nuclear power both through enactment of the Price-Anderson Act, which limits a utility's liability in the event of an accident, and through the establishment of what is now the Nuclear Regulatory Commission (NRC), which approves the plans, monitors the construction, and issues construction permits and operating licenses for all commercial reactors in the U.S. with a specific view toward safety.

EVOLVING REGULATORY PHILOSOPHY

While most privately regulated utility investments are conceptually similar to publicly provided capital infrastructure, and I assert that the proper management procedures for planning and maintaining both are strikingly similar, the obvious difference is that there is no private ownership equity associated with publicly funded projects. As a consequence, the major philosophical debate in private utility regulation has arisen over the approval of prices and how *high* those prices must be in order to afford the owners of the capital an opportunity to earn a fair return on their investment. By contrast, the debate over publicly provided infrastructure usually focuses on how *low* the price can be made so that all potential buyers are happy. The net effect of those different standards is obvious.

Electric service in New York State may appear expensive, but it is reliable. By comparison, New York subway fares have been kept low, but the system is a disaster. In fact, one of the many reasons electric rates in New York City are so high is because the legislature approved a special profits tax, which falls heavily on utilities, in order to keep subway fares low. I know many public service commissioners would welcome the reciprocal opportunity of postponing a ConEd rate increase by hiking subway fares.

The legal precedents on private utility pricing go back to an 1898 case, *Smyth v. Ames*,[14] which established the principle that regulators must set a price that affords the investors an opportunity to earn a fair return on the market value of their property. After years of litigation to interpret the meaning of "fair return" and "market value," the Supreme Court in *Bluefield Waterworks and Improvement Co. v. The PSC of West Virginia* declared that the prices should be set so that the rate of return would permit firms to raise required new capital and to earn a return on their investment that was equivalent to that of firms experiencing similar levels of risk.[15] Because this ruling did not nail down a formula to determine fair market value, in a 1944 case, *The Federal Power Commission v. Hope Natural Gas*,[16] the Supreme Court maintained that all that was required of regulation was that it led to a reasonable result, independent of the computational formula. This decision in effect endorsed the application of straight line depreciation to the embedded book cost of a facility in setting a revenue requirement for the determination of electric rates, so long as the outcome was reasonable.

While that 1944 decision may have ratified computational simplicity in the rate-making process, thereby easing the burden on accountants and lawyers, and may in fact also have led to reasonable results in the noninflationary period of the 1930s and 1940s; in light of the inflationary experience of the recent decade, it has bred nothing but trouble. As an example, were we to revert to the *Smyth v. Ames* nineteenth-century standard, perhaps some of the public's concern over the Shoreham nuclear plant on Long Island would disappear. After all, what is the current *market* value of Shoreham? What would someone be willing to pay to buy the plant from the Long Island Lighting Company? Answer: probably not much. So, by that standard, the PSC would be required to place very little of the plant's actual cost into rate base initially.[17]

More recently, through the Public Utility Regulatory Policy Act of 1978 (PURPA), the federal government has opened the electric industry's door slightly to competition in generation from small hydroelectric plants, from power generated from renewable sources or wastes, and from cogeneration. Electricity generated by plants fueled by hydro-power or renewable sources, or cogeneration, must be purchased by utilities at their own avoided cost. And in December 1983, the FERC paved the way for a competitive experiment in the Southwest among large, utility-built central-station generating plants. So in one respect, the industry's regulatory structure seems to be reverting to its nonregulated status of a century ago.

TOPICS FOR THE FUTURE

The real concern today about electricity supply involves its potential importance for the future, and the reason for this brief restrospective is to set the stage for the subsequent prospective analyses. And while our forecasting skills have improved, our historical experience is nevertheless humbling. As an example, as electricity demand growth began to stagnate in the 1970s, it became increasingly apparent that if we were ever to project future demands reasonably, a much better understanding was required of income and price elasticity responses. Even after those mechanisms began to be understood in the mid-1970s, a time when the country and New York were undergoing substantial economic retrenchment, it became far more important to understand and predict the economic future of New York State if a reasonable estimate of electricity demand were to be provided. Therefore, a major focus of this volume concerns the interrelationship between the economy and the supply of electricity, but it also considers the impacts that environmental, health and financial concerns—all linked through public opinion and the regulatory process—might have on the future of the industry.

NOTES

1. See H. C. Passer, *The Electrical Manufacturers* (Cambridge, Mass: Harvard University Press, 1953), p. 112.

2. Ibid., p. 139.

3. See T. Bernstein, "A Grand Success," *IEEE Spectrum* (February 1973):55.

4. See E. D. Adams, *Niagara Power*, vol. 2 (Niagara Falls: Niagara Power Company, 1972), p. 39.

5. New York, *Times*, September 18, 1927, sec. 7, p. 16.

6. "70,000 H.P. Station," *Scientific American* (January 13, 1900).

7. See J. Bauer and N. Gold, *The Electric Power Industry* (New York: Harper and Row, 1939).

8. "From Shafts to Wires," *Journal of Economic History* 63, no. 2 (June 1983):347–72

9. See Passer, *Electrical Manufacturers*, p. 293.

10. See New York State Public Service Commission, *Utility Regulatory Bodies in New York State 1855–1953* (Albany: New York State Public Service Commission, 1953), p. 19.

11. See N. A. Turkish, "PASNY and Its Policies," Master's thesis, Cornell University, 1960, p. 4.

12. See Paul D. Mazzarella, "Critique of Power Plant Siting under Article VIII of the New York State Public Service Law," Master's thesis, Cornell University, 1979, p.5.

13. See R. D. Baum, *Federal Power Commission and State Utility Regulation* (Washington, D.C.: American Council on Public Affairs, 1942), p. 35.

14. 169 U.S. 466 (1898).

15. 262 U.S. 679 (1923).

16. 320 U.S. 591, 605 (1944).

17. Conversely, using the market value, rate base standard would mean that many upstate utilities would receive cash flows based upon $1,200/kW, current coal-fired plant costs, not their historic $100/kW embedded costs. And by the same standard, the NYPA hydroelectric power should sell for 6–8¢/kWh, not at 5 mills to 2 cents, as in some cases.

2

The Energy Crisis
Ten Years Later

Lawrence R. Klein

THE MEANING OF THE "CRISIS"

Riding the Metroliner to Washington, D.C., one evening in the waning weeks of 1973, I heard fellow passengers talking about the energy situation and what was then being done about it. A self-assured disputant declared that there was no shortage. The whole crisis atmosphere was created by public authorities, for there was plenty of energy available, even in the form of gasoline. The authorities simply wanted to exert control over people's lives or were ignorant of the true situation.

In some respects, the disputant was right. There were enough BTUs available for use over a long time into the future. He was also probably right that we could have kept consuming refined oil products in our usual way and we would not have run out in the foreseeable future. After all, the embargo, imperfect as it was, lasted only about six months. We would not have exhausted available supplies over that period.

The arguing fellow passenger did not *perceive* that a crisis existed, yet the crisis was real; it *did* exist because people perceived that all was not right and that it was prudent to consume energy in a constrained way. We could not be sure, immediately, of the duration or effectiveness of the embargo. Actually, it did not deplete our reserve resources, in the sense that our consumption in that short period did not effectively hurt our reserves, and it has been relatively easy to add to reserves since that time.

The embargo, as long as it lasted, and the subsequent changes in the terms of trade between producers and consumers constituted the crisis. The new relative prices for energy were set at such a high multiple of their previous values that we could not easily or quickly adjust to the new

18

situation. That constitutes a crisis. Timing and perceptions were the forceful factors that made a crisis out of the situation.

The feeling that we were in a crisis prompted researchers to become knowledgeable about energy—to think in energy terms, to learn the energy vocabulary, and to bring energy sectors into models of the economic process. As an economist working on quantitative studies of the moment, I was forced, with my colleagues, to adapt our econometric models in a "quick-and-dirty" way in order to get some plausible, usable results in the short run.[1] But, step by step, we introduced full-blown energy sectors into our models. Now we more fully appreciate the role of energy in the economy, and the energy sectors are here to stay. We may well not return to the crisis atmosphere of 1973–74, but we will keep the energy details of our models intact.

When we periodically—say, every month or so—set out on new analyses of the economic situation, we systematically thought first about various energy magnitudes, to be exogenously fixed in order to simulate our systems. We estimated paths for prices, for production quotas, and for types of energy available. These were considered to be among the most important inputs for our models every time we assessed them for new findings.

The energy problem was looked at in many dimensions. This problem gave rise to complications for:

- production
- cost
- inflation
- lifestyle
- balance of payments
- exchange rates
- sales of associated goods (e.g., cars, houses)

Ten years ago the energy constraints on economic performance were quite significant. Even if we feel quite relaxed about available energy now and do not attach so much importance to energy inputs for our models, that does not mean that we were behaving inappropriately during the past decade, and it does not mean that the problem was not serious.

Our perceptions made the energy problem serious, particularly before we were able to see progress made in overcoming potential dangers. Also, we could not tell how long the crisis atmosphere could be maintained. The problem lasted for nearly one decade and it was quite right to take major steps to deal with anything that was perceived to be a problem for so long.

One of the biggest issues associated with the oil crisis was its impact on world trade and payments, particularly the international payments of various countries. Early on, it was perceived that major oil exporters would accumulate large financial reserves, while importing countries would have large trade deficits to be financed. Just as some people feared that industry would grind to a halt for lack of oil, many feared that the largest oil producers in the Middle East would come to own the bulk of the world's assets. Obviously, the most feared of these dire circumstances did not come to pass, either in the spheres of production or finance, but it is extremely useful to look back on the way the oil crisis was financed, that is, on the way petrol dollars were recycled.

Despite inflation and dollar depreciation, major oil producers accumulated large reserve balances. Some of these funds were invested in real assets, but a great deal went into equities, bonds, notes, bills, and bank accounts. The large sums that went to banks swelled the reserves of several large money center banks operating in the international theater. There is a dispute about whether total bank reserves increased or whether they were just redistributed. I notice large increases in official reserves after 1973, and these can form the basis for a general deposit expansion. There was an expansion for a few years right after world monetary reform, when the fixed Bretton Woods' parities were dropped in 1969. Until 1982, reserves grew on a high scale every year, except possibly in 1973 (see Table 2.1).

Whether or not total reserves rose on a world scale after 1973, it is a strategic fact that large international banks had fresh reserves in the form of the profits of OPEC and other oil exporters. It can be said that banks abhor idle reserves. They started to (try to) turn these reserves into productive assets and encourage developing countries, among others, to obligate themselves through more bank borrowing. It is well known that

**TABLE 2.1 Total Reserves Minus Gold
(in Billions of SDRs at End of Year)**

Year	World Total	Change	Year	World Total	Change
1968	39.1	3.9	1976	188.1	28.0
1969	39.8	0.7	1977	229.8	41.7
1970	56.3	16.5	1978	246.9	17.1
1971	87.8	31.5	1979	274.0	27.1
1972	111.8	24.0	1980	325.4	51.4
1973	117.8	6.0	1981	342.5	17.1
1974	145.5	27.7	1982	337.9	−4.6
1975	160.1	14.6			

Source: International Financial Statistics, International Monetary Fund.
SDRs: Special Drawing Rights.

loans were pushed onto the books of developing countries and this is when the trouble began. Many of these loans were at variable interest rates; so with the sag in oil and commodity prices, together with the large rise in interest rates, many developing countries, both oil exporters and importers, found themselves with unmanageable debt burdens. The worldwide recession of 1981–83 made the situation even worse for developing countries. These years mark the beginning date of the world debt crisis, which continues to threaten the stability and even existence of the world monetary system. We thus see a fairly direct link between the world's energy crisis and the present world debt situation.

Recycling problems hit not only developing countries who had borrowed heavily in international financial markets but also domestic banks. The failure of Penn Square Bank and with it the acute problems of Seattle First National and Continental Illinois are simply different aspects of the same problem. Sagging or falling oil prices since 1980 have made domestic oil operations risky, and those who borrowed heavily for financing their activities have often been unable to pay. The Penn Square failure, centered around Enid, Oklahoma, was analogous to the financial problems of Mexico, Venezuela, Nigeria, and Indonesia.

Thus, the oil crisis has been far-reaching, and it has probably not yet run its course. Although the worst possible situations that were feared from the start, right after the embargo was announced, have not taken place, the events that have occurred are serious and still command a great deal of attention even though other aspects of the crisis have faded into the background.

RESPONSES TO THE CRISIS

Assuming that we now know, at least in a general sense, what the meaning is of the term *oil crisis*, what has the economic response been? As I have indicated, the response has not yet been resolved with respect to financial complications, but some very significant steps have already been taken with respect to the physical crisis of supply.

It is very important to realize that economic forces relieved the extreme pressures of the crisis by

- stimulating new production and the search for more supplies of energy—especially oil and gas;
- substituting new energy sources for the traditional patterns (this has involved inter-fuel substitution); and
- inducing conservation through car redesign, changing the size of the units in the national fleet, redesigning new aircraft, setting highway speed limits, lowering thermostats.

To a large extent, relative prices were instrumental in bringing about conservationist tendencies. Direct allocations and controls can also achieve highly favorable results, but the broadest changes have been a result of price factors.

Some people thought that energy demand was completely price inelastic (insensitive to price changes) as far back as 1974, and others thought that the elasticity coefficient was tiny. As a result of good economic research and good, though less formal, empiricism about the world we live in, it has become evident that price elasticity exceeds 0.1 and may be significantly greater than 0.5, but the higher elasticity estimates are longer-run figures, needing two or more years to work themselves out. For the short run, price elasticity is more likely to remain at 0.1–0.2. It may seem that these are quite small numbers, creating much economic uncertainty about handling the problem. The elasticities have, in fact, been pinned down to these stated ranges, and it turns out that small elasticities of about 0.2 make a great difference in the resolution of the problem. Were the true elasticity to be zero, we would be having a great deal more trouble than we do at present.

As prices went up, people became much more mindful of conservation and watched their energy consumption. In some cases they switched from oil to gas, coal, or other fuel sources. In other cases they simply reduced consumption or introduced new capital items to improve the use factor. For example, improved home insulation is an instance of reducing the amount of energy used while achieving the same comfort level.

Price effects did their work on the side of supply as well as of demand. More drilling rigs went into operation and more oil and gas were found. On a world scale, Mexico became a major exporter; the North Sea was more fully exploited by Britain and Norway; smaller fields were discovered in a number of countries. Argentina, Australia, Egypt, and Malaysia became more self-sufficient in oil. Others also became practically self-sufficient. Brazil continued to follow its own distinctive line of supply enhancement by promoting an alcohol program for its car fleet and other uses.

Within the United States, the shift in fuel source for energy supply has been remarkable. At the onset of the crisis, in 1973, we derived 35.04 quad BTUs from petroleum, 23.17 from natural gas, and 12.92 from coal. By 1982, these figures were 29.56, 18.91 and 15.89. Coal use expanded by almost 25 percent, while oil and gas use declined. There was also a small gain in the use of non-fossil fuel electricity (hydro, geothermal, solar, and nuclear). At the same time the economy grew, despite some intervening recessions. The principal factor was an improvement in the use of energy, that is, a general gain in efficiency. BTUs per unit of real gross national product (GNP) have fallen steadily since 1973. There is some cyclical effect in the figures on per capita consumption, but the trend is down. As for the energy efficiency of producing the GNP, there is clearly an improvement[2] (see Table 2.2).

At the same time that use patterns, both in production and consumption, have changed, we have come to a better stock position. The strategic reserve is gradually being filled—toward targets of a certain number of days' supply. When Mexico was in a severe payments crisis because of the debt service burden, the U.S. bought unusually large amounts of oil for the reserve. This was good policy for both countries, given the circumstances of the moment.

RESULTS OF THE ENERGY RESPONSES

It is well known that we in the U.S. and the world at large went through an unfortunate decade during the 1970s, the same decade that contained two oil price shocks together with some actual and some threatened supply limitations. It is always difficult to establish causation in economics, but the energy shocks were so large and the related events were so close in timing that it is difficult to ignore the evidence of various energy-induced effects during this decade.

The main economic features of the decade of the 1970s were a growth slowdown, with recessions; a rise in inflation; a decline in the rate of productivity; and a threat to world financial stability.

Growth Slowdown

Before the onset of the embargo and the subsequent increase in oil prices, it was noted that the increasing dependence of the United States on oil imports—for cars, electric utilities, space heating, jet aircraft, and so on—would impose a burdensome deficit on our international payments. We were forced to project our Wharton Model simulations more cautiously as early as 1970 in order to hold down the unfavorable trade balance. This problem worsened after the change in the terms of trade in 1973–74, and the U.S. growth rate was marked down, by consensus, from 4 percent to 3 percent. The same was true of Western Europe. The Japanese downshifting was even more drastic, from 10 percent to 5 percent. These perceptions of growth potential prevailed all during the 1970s and remain as the consensus view for the 1980s, but some people are now looking toward a revival of stronger growth. They are not part of the consensus. Those among us who want to rely on strong growth for helping to solve our federal deficit problem must look to 4 percent growth or better for the rest of this decade. It is possible but not easy to achieve.

TABLE 2.2 U.S. Energy Input and per Capita Energy Consumption

	1973	1974	1975	1976	1977	1978	1979	1980	1981	1982	1983
BTUs×10³											
GNP 1972$	59.17	58.16	57.23	57.23	55.64	54.24	53.50	51.46	48.90	47.85	46.03
BTUs×10⁶											
Person	350.2	338.9	326.4	340.8	346.1	350.6	350.3	333.3	320.8	303.7	300.0

Source: Wharton Econometric Forecasting Associates.

Not only was the growth trend restrained during the 1970s, there were also two recessions that followed like clockwork in the wake of each oil price shock. It is because of this timing that I feel comfortable in associating the recessions with the oil price shocks, despite the fact that other events related to the recession events were occurring at the same time (see Table 2.3).

With an output slowdown, recessions, and strong growth of the labor force, it is no wonder that unemployment rose—to heights that we had not experienced since the dreaded days of the Great Depression.

It was unusual to have inflation with recession or rising prices and rising unemployment. The Wharton Model's simulation properties exhibited similar results, however. When the model was shocked with an external price disturbance in either food or fuel, we got a stagflation relationship rather than a trade-off relationship, namely, more unemployment and more inflation together.[3]

Rise in Inflation

Again the timing was remarkable. As soon as oil prices exploded, the main price indexes (consumer price index, GNP deflator, or producer price index) shot upward. The coincidences were too striking to ignore. Some economists argued that we merely had a change in relative prices, not a case of classical inflation, but the broad index averages moved upward, and that is what matters. The oil shocks were parts of a rolling series of relative price changes, and this rolling pattern had brought about general inflation. Also the deterioration of the U.S. dollar in this situation worsened the inflationary pressure, as did the built-in tendencies of wages to follow the cost of living upwards.

Decline in Productivity Rate

The productivity slowdown has puzzled many economists. I think that it is very much a consequence of our energy problems. During the 1970s we spent a great deal of effort and time in becoming energy efficient. This result is evident from the data in Table 2.2. In concrete terms, we changed the car fleet; we shifted from burning oil to burning coal in many electric utilities; and we produced new kinds of aircraft. Some of these changes were linked to environmental protection, as in the use of scrubbers for electric utilities. It took a full decade to change the fleet, to adapt utilities, and to make similar changes with respect to energy use, which held back conventional measures of productivity. Energy productivity improved, but labor productivity did not: it fell in some years and slowed in others. The productivity slowdown made the inflation problem worse.

TABLE 2.3 Inflation and Growth of GNP versus Oil Price Change, 1973–1983 (Percent Change)

	1973	1974	1975	1976	1977	1978	1979	1980	1981	1982	1983
GNP (1972$)	5.8	- 0.6	- 1.2	5.4	5.5	5.0	2.8	- 0.3	2.5	- 2.1	3.7
CPI-Urban	6.2	11.0	9.1	5.8	6.5	7.7	11.3	13.5	10.4	6.1	3.2
Crude oil import price	22.9	243.5	3.4	7.2	6.7	1.3	39.1	67.8	12.0	- 8.1	- 11.9

Source: Wharton Econometric Forecasting Associates.
CPI-Urban: All items, all urban consumers.

Threat to World Financial Stability

It may seem far-fetched to blame the world's financial problems on the energy problem, but the recycling of petrol dollars, as explained already, was accomplished by pushing high interest loans both on our own banking system, as in the Penn Square case, and on developing countries. Reckless public financing measures and inflation pushed interest rates up, and the financial crisis became fully apparent. The worst manifestations were in the developing countries, where an about-face in economic growth was suddenly imposed by the beleaguered financial system trying to recycle petrol dollars.

Outcome of Oil Crisis

We have been through many crises; some are not yet resolved and others may reveal themselves in a surprising way, but in many respects we are less vulnerable for having survived. The various points that have been stressed in this chapter—namely, conservation, fuel efficiency, a shift from oil to other fuels, stockpiles, and less fear of OPEC as an absolute factor in the world oil market—work jointly to make us less vulnerable. In many respects, because of these points, we as a nation will probably be better positioned to face oil shortages in the future, either as a result of an interruption in supply or because of other forces that cause prices to rise. Our *perception* of the problem may be different from our earlier one, and we may not have the severe reactions that we had in 1973–74 or 1979–80. Generally speaking, we may face new oil shocks in a more relaxed manner. We may react in a manner that would suggest wonderment about why we were so disturbed about the oil price shocks of the 1970s.

The market problem of today is not one of price pressure upwards, but one of weak prices. If price rises in the presumed "tight" markets of 1973–74 and 1979–80 caused domestic stagflation, in extreme measure, can we expect opposite reactions in symmetrical amounts when prices fall? The first round of oil price cuts in 1980–83 already throws some light on an appropriate answer to that question. The direct impact of oil (and other commodity) price leveling or dropping from 1980 prices was instrumental in bringing down the rate of inflation. Monetary control and recession were major factors in 1981–82, but the process began with a reversal of oil price movements.

By an indirect route, the surge in the dollar, which is partly related to the flight of capital to the U.S., which, in turn, was partly a result of debt problems and recycling of petrol dollars, contributed further to the decrease of U.S. inflation. This was not so for inflation in other oil-importing countries.

Lower inflation and recovery from recession were both helped in the U.S. by sagging oil prices. This is not to deny the influence of many other factors, but the reverse effects of falling prices did their bit, and if prices continue to come down in 1984–85, they should bring additional benefits. Falling oil prices will make the gains in fighting inflation all the more secure and give encouragement to the Federal Reserve authorities to make monetary policy even more accommodating.

But the results are not expected to be completely symmetrical. Oil price drops in the context of 1984 are not the precise mirror image of the price increases that occurred under the conditions of 1973–74 and 1979–80. When we run our econometric models through reverse price shocks, we do find nearly symmetrical quantitative results. The main asymmetry, however, is the impact on the export earnings of Mexico, Venezuela, and other major oil producers. We have a great stake and interest in the economic well-being of these oil-exporting nations. For world financial stability, as well as pure friendship, our gains from declining oil prices would have to be used in part to help Mexico, either through financial concessions with regard to debt servicing or through policies that help to maintain export earnings.

A consensus view is that oil-importing countries should take their gains that result from lower oil prices and make direct contributions to developing countries who would be experiencing international payment difficulties. The industrial countries would realize a net gain, even though the outcome would not be symmetrical in comparison with the case of the price rise.

Our economic concern would be not only for some developing countries who are oil exporters, but also for Great Britain. There could be major trouble for the pound sterling, for inflation, and the general health of the British economy. Problems are not expected to occur with obvious severity for price changes of about $1.00 or $2.00 a barrel, but if there is a more substantial price drop of $5.00 or more, the side effects could be very serious, indeed, and would require explicit action by the U.S.

Within the U.S. there would be gainers and losers. Some areas and some industries benefit directly from high and rising prices. By and large, the northeastern industrial areas are large importers and consumers of oil. They would stand to benefit, both directly and indirectly, from falling world oil prices, since other fuel prices would move in sympathy with crude oil prices. Undoubtedly, New York State and especially electricity users or producers would stand to gain. For the country as a whole, just as for the world, the benefits should outweigh the losses.

PROSPECTS FOR NEW YORK STATE

New York, like other states in the Northeast, did not fare well during the 1970s. This is true both in an absolute sense and in relation to the country as a whole. It is not difficult to recall the fiscal crisis in New York City and the associated involvement of the entire state's finances. For some time, at least, that has been a thing of the past.

New York is a comparatively rich state. The per capita disposable income in 1972 dollars is $6,030, versus just $4,919 for the U.S. as a whole. By contrast the unemployment rate is slightly higher, averaging about 7.75 percent during 1984, while the corresponding national figure is 7.34. Youth unemployment in New York City must surely be a contributing factor to the discrepancy.

Also, the high cost of living in New York City is well known. This does not mean that inflation is significantly higher than in the country at large, only that the price *level* is higher.

Like so many northeastern states, New York lags behind the nation in population growth. This reflects the Snowbelt/Sunbelt rivalry. In New York, the rate of population expansion for the coming decade is expected to be only 0.2 percent, while the corresponding figure is just under 1.0 percent for the nation as a whole.

A useful comparison, past/present/future, for the state and nation is given in Table 2.4. In both output and employment growth, New York has lagged behind the rest of the nation as a whole, despite the fact that incomes are much higher in New York. For the coming decade this comparative situation is expected to persist, but a catch-up period is in sight for the 1990s.

But productivity growth looks better for both past and future in New York. This explains, in part, the higher per capita income position of New York State. New York will not be the national leader in high technology, but it will be well represented. As a consequence, employment opportunities should continue to be good in new fields, such as electronics, in New York State, where some of the industry giants are prominently located (General Electric, IBM, Kodak, and Xerox, to name some of the best known).

What can we say about energy trends on the basis of this thumbnail economic profile? In the first place, energy needs will continue to grow, especially in line with the projection that economic growth will pick up to a rate above the national average. It should be noted as well that both New York and national estimates are on the conservative side. There may well be a significant upside risk to this forecast, especially if the new technologies continue to catch on as well as they have. Electricity generation

TABLE 2.4 Output, Employment, and Productivity: New York State and the Nation (Percent Change)

	1974–1983	1984–1993	1994–2003
Output			
New York	0.9	2.3	2.8
U.S.	2.0	2.9	2.4
Employment			
New York	0.2	1.2	1.2
U.S.	1.8	1.8	1.1
Productivity			
New York	0.7	1.1	1.6
U.S.	0.2	1.1	1.3

Source: Wharton Econometric Forecasting Associates.

grew from 108,995 million kWh in 1974 to 122,371 million kWh in 1983, which was a very slow period for New York State. With the growth rate more than double for the coming decade, we can look for much faster expansion in the electrical sector during the next ten years.

The importance of coal will continue to grow, as it has since 1974. Natural gas and nuclear power also expanded greatly over the past decade. We can look for continued expansion in the use of natural gas for electricity generation, but the future of nuclear power is very uncertain. It is not likely to be an expansion area for some time, given public attitudes toward its use and development.

A feature of electricity generation in New York has been the decreasing use of petroleum products for fuel. Their use grew rapidly during the 1960s and reversed itself during the 1970s, in adjustment to the energy crises. The pattern of use in New York State reflects what has been accomplished throughout the country by way of inter-fuel substitution and conservation.

Just as a feeling that crisis situations are things of the past—not forever but for now—prevails in the nation at large, so should it prevail in New York.

NOTES

1. L. R. Klein, "Supply Constraints in Demand Oriented Systems; An Interpretation of the Oil Crisis," *Zeitschrift für Nationalökonomie* 34 (1974):45–46.

2. The rise in the ratio of electricity to GNP can be attributed to economic reasons. Gross output originating in electricity production grew in real terms by about 35 percent between 1973 and 1983, while real GNP went up by only 22 percent in the same period. Units of kWhs generated per unit of GNP did not rise between 1973 and 1983; this ratio remained practically constant. Electricity prices rose by a little more than 200 percent (i.e., tripled) in the decade after 1973, while petroleum and natural gas prices went up by much larger multiples.

3. Bert G. Hickman and L. R. Klein, "Wage-Price Behavior in the National Models of Project LINK," *American Economic Review* 74 (May 1984):150–54.

3

The Long-Run Security of the Energy Supply

James R. Schlesinger

With respect to its energy supplies, the United States regularly goes through cycles of complacency and panic. We are now well advanced into another period of complacency. We can be quite confident that this period of complacency will be succeeded, as it has been in the past, by a renewal of panic. Indeed, one can say with similar confidence that this period of complacency normally lays the seeds for the subsequent energy crisis.

All that I am required to do in this chapter is to forecast the future. One of the wisest American philosophers, Samuel Goldwyn, said "Prediction is difficult—especially about the future." How can we best deal with the future? As an eighteenth century political writer once said: "The best way to be a prophet is to have a good memory." So, as we look toward the future, we are obliged to recall the past.

To address the topic of New York's future power supply, I cannot treat either power supply or the state itself in isolation. Instead, I must cast my net far more widely; I must deal with other forms of energy and a much wider geographical area.

The demand for electric power in the U.S. can be determined with reasonable accuracy if one knows just three variables: the price of oil, the gross national product, and the rate of interest. Together these three will determine the conditions facing the electric power industry. Needless to say, this equation represents something of an oversimplification. Nonetheless, it is one that will be useful for our purposes.

The price of oil is critical because oil is the fuel that substantially determines the prices of other fuels. Any interruption of oil supply will be reflected in its price. As we know, oil supply has now become the pre-eminent security-of-supply problem, a fact that primarily reflects the worrisome dependence of the free world on the Persian Gulf—politically the most volatile region of the world. That region also happens, unfortunately, to be perilously close to the concentration of Soviet power.

The rate of interest is easily established—under full employment conditions—if one examines the state of the federal budget. The immense deficit in the federal budget, which I shall not discuss here at any length, has reduced the supply of domestically provided capital under full employment. One consequence is that the utility industry now faces real interest rates that are staggeringly high by historical standards. In the short run, the impact of the federal deficit on the availability of capital has been offset by an astounding inflow of foreign capital—now amounting to $100 billion a year. But that inflow itself is dependent upon an overvalued U.S. dollar. At some point in the future the dollar will decline and the inflow of foreign capital will presumably cease. At that time we will face the true consequences of a deficit in the federal budget amounting to 5 percent of the gross national product—an amount almost equal to the totality of personal savings in this country.

The third key element in determining the condition of the utility industry is the gross national product. Depending on your preferences, you can choose one of several projections of the GNP. If, for example, you believe in unfettered supply-side economics, the future is rosy, but *electric power supply will have to grow very rapidly*. If you believe in the Reagan administration's current projections—somewhat less rosy, but still a robust 4 percent—problems of power supply will not arise until around the end of the decade. I shall return to this subject.

THE SUPPLY OF OIL

Let me now, however, deal more carefully with one of these three key items: the price of oil, which has become so critical. In recent years oil markets have acquired a new characteristic that did not exist earlier: a high degree of volatility. This new volatility in the oil market reflects, of course, the ebb and flow of demand. Fluctuating demand itself is, in turn, a reflection of drastically altered attitudes toward inventories.

Since the price of oil has risen substantially over its pre-1973/74 price, the carrying costs for the industry of its inventory have risen astronomically—from roughly $8 billion to over $120 billion. Consequently, the industry now finds it economical to adjust inventory levels, depending on its belief in whether the price will rise or fall in the near term. If prices are expected to fall or even to remain stable, inventories will be reduced. In periods of rising prices they will surge. Such a surge of inventories is typically what underlies the appearance of a shortage— and what turns out more accurately to be a panic over supply.

Inventory volatility is not the only type of volatility that we have observed. We may also observe volatility in energy prices themselves as well as an entirely new volatility in exchange rates, especially important in the energy market since oil prices are denoted in dollars.

In recent years, for example, despite the "soft" oil market the price of oil has continued to rise in real terms in Western Europe; this is because the American dollar has risen. The rise in the American dollar in terms of the local currencies has more than offset the relatively small decline in the official selling price for oil. Moreover, the rise in the price of energy, driven by the movement of exchange rates, has served to intensify Europe's economic woes.

In dealing with the energy problem, we should recognize that we are faced with a fundamental irony: those things that are beneficial and comforting in the short run tend to contribute to and intensify the longer-run problem. Conversely, those things desirable in the long run tend to impose short-term costs. Reinforcing the cycle of complacency and panic is the psychological fact that in periods of complacency we are especially reluctant to incur the short-run costs that might ease our longer-term energy problem.

Of late much has been written about the reputed collapse of OPEC, and the current, "soft" condition of the oil market. One should bear in mind that a collapse of OPEC at this time would inevitably intensify the longer-term energy problem, even though, in the short run, a decline in oil prices would certainly be widely welcomed. A sharp drop would undermine much of the effort to develop alternative sources of supply. That would include, notably, much of the free world, where drilling costs would then become uneconomically high.

Although there is significant excess production capacity at the present time—amounting to perhaps 8 million barrels a day (mbd)—all of it is located in OPEC. We should also recognize the painful reality that, despite a worldwide recession, despite continued efforts at conservation, and despite a doubling, in real terms, of oil prices in recent years, the total decline in the demand for oil in the free world has only been about 15 percent.

We are now in the process of reversing this earlier decline. Oil demand is growing once again at the rate of approximately 1 mbd every year, worldwide. Both the reality of recent growth in imports and the Department of Energy (DOE) projections for the U.S. for the near-term future underscore the prospective occurrence of an oil supply crisis sometime in the 1990s. From 1983 to 1984 oil imports into the U.S. appear to have risen by 1 mbd, from 4.5 to 5.5 mbd. And as we look forward to 1985, the projection by the DOE is for imports of 6.2 mbd. By 1995, the DOE projects 9 mbd of imports. That is a reflection not only of rising domestic demand but also of the projected decline in America's capacity to produce crude oil—a capacity that is expected to fall by approximately 1 mbd by the early 1990s.

Most interestingly, included in that prospective 9 mbd of imports is an increase in the projected use of oil by power plants, from approximately 600,000 barrels a day to 1.6 mbd, in other words, a rise of 1 mbd. This rise reflects in part the expanded use of oil-fired combustion turbines, as well as the *expected* return to service of the oil-fired generating facilities that have temporarily been retired because of the substantial reserve generating capacity presently available. (One should add that reference to this capacity as *temporarily* mothballed is an act of faith; we can only hope that all of it can actually be fired up.) I also might add that the lagging construction of additional alternative power generating capacity means that in the 1990s we shall be forced back into greater dependency on oil-fired plants.

Nonetheless, for the time being, the Western world has adequate capacity to cope even with a supply interruption in a major producing country. It could not, of course, effectively handle the closure of the Persian Gulf for an extended period of time or, even worse, a Soviet takeover of that region. But today no *single* source of supply is critical to the world oil market. That is likely to change.

We have already passed the point at which the growth of non-OPEC sources of supply is likely to alleviate the problem of dependence on OPEC. Indeed, dependence on OPEC is virtually certain to rise. As I mentioned earlier, American oil production is likely to decline by 1 mbd by the early 1990s. In the British sector of the North Sea, which will top out in 1985 at 2.5 mbd, we are just now hitting the decline curve. By the early 1990s production is anticipated to decrease to 1.4 mbd. The Soviet Union, which has been eager to exchange oil for hard currency, is likely to decline as a source of supply for the West: Soviet production peaked out this year at something over 12 mbd and will gradually decline. That famous—or rather, notorious—1977 analysis by the CIA, which projected a sharp decline in the oil the Soviets could make available outside of the Soviet Union and outside of the Soviet bloc, was clearly premature, in part because the Soviets took offsetting actions. Nonetheless the analysis was basically correct, even though the date of the start of the decline has occurred some five years later than initial CIA projections.

The outcome, then, will be a decline in availability from the Soviet quarter. Canadian production will also be declining, and early in the next decade Canada once again will be forced to turn to external sources of supply. Moreover, a number of the OPEC nations will find themselves hard pressed to maintain capacity.

Inevitably, the free world will again become more dependent than ever upon OPEC—and more dependent upon the Persian Gulf. Collectively, the OPEC nations control well over 70 percent of the free world's oil reserves. We are now depleting oil reserves more rapidly outside of OPEC than within OPEC. The inevitable consequence will be a rising dependence on the Persian Gulf—at just the moment when the political stability of the Persian Gulf will likely have become even less assured than at present.

The Middle East is without question the most volatile region of the world. It is akin to the Balkans before World War I, a potential tinderbox for international rivalries. Within the region of the Gulf the Soviet Union has gradually become the dominant military power. It remains so today. Yet the U.S. has pledged to protect the nations of the Gulf against the intrusion of outside military power. At the moment, however, the military deterrent that the United States can provide in the region may be described most charitably as a low-confidence one.

The position of the U.S. in the region has been slipping steadily over the course of the last fifteen years. With the fall of the Shah of Iran, it declined precipitously. Whatever his real or supposed vices in the area of civil rights, the Shah was the linchpin of American policy. His fall was a catastrophic blow to our geopolitical position. The politics of the region seem to be growing steadily more precarious. Consequently, it would seem prudent for the West *not* to increase its dependence on that region. Yet that is precisely what we are going to do.

In recent years Americans have been lulled into complacency by a declining oil demand, driven partly by higher prices and partly by recession. But in the years ahead as oil demand rises and as non-OPEC sources of supply shrink, we shall once again be driven to greater dependence on OPEC. Because of our current complacency, we are now sowing the seeds of the next energy crisis. By the early 1990s we will most likely see another supply interruption and another crisis, given the instability of the politics of the region.

What will be the consequences for energy supply in the U.S.? The first point to bear in mind is that an increase in oil prices tends to raise the prices of competing fuels. Natural gas prices move up more or less parallel to the increase in oil prices. Coal prices move up somewhat, but proportionately less than the increase in oil prices. The price of electric power is influenced least of all. In the aggregate, it is not raised in the same proportion as oil prices—even for oil-fired plants. The effect of any increase in oil prices is to improve the competitive position of all substitutes for oil. It is important to recognize that such an increase most sharply improves the competitive position of electric power.

I have already suggested that we are likely to face tight oil markets in the early 1990s—with the likely possibility of a new supply interruption that will significantly and suddenly reinforce the upward pressure on oil prices arising from those tight markets. If this were to occur, there would be powerful pressure to substitute electric power for oil—at just that point when the country is likely to be experiencing problems with overall generating capacity. So now, at long last, let me turn to electric power demand and supply.

ELECTRICITY SUPPLY AND DEMAND

For reasons that I shall spell out in a moment, the utility industry has been strongly motivated in recent years to minimize—indeed, to understate—the projected increase in electric power demand. The industry as a whole has been projecting demand for electric power to rise at 2.8 percent a year. Meanwhile, in 1984 alone electric power demand rose by some 7 percent, the highest rate of growth for electric power demand that we have observed in a decade or more. As a general estimate I would suggest that the demand for electric power will grow at perhaps 0.5 percent more than the growth of the gross national product.

The administration is currently projecting a steady real increase in the gross national product of some 4.0 percent a year, which suggests that the demand of electric power might rise at 4.5 percent a year. Thus, there is a notable gap between the current projections by the industry and any projections based on the relationship between the GNP and electric power demand. That gap can only be explained by the happenings in recent years on the side of regulation, so let me finally raise the question of electric power *supply*.

At the moment we are comforted by the apparent excess of capacity. But that excess of capacity would melt away reasonably rapidly, if power demand grows at 4.0 or 4.5 percent. And the prospective growth of additional capacity will ultimately reflect two things: the rate of interest and the regulatory environment.

There has been an understandable concern about the rate of interest throughout the private sector. As Helmut Schmidt, until recently the West German chancellor, has repeatedly (and accusingly) stated: the real rate of interest is now at the highest point since the time of Jesus Christ. In large measure, that is a reflection of expectations regarding inflation and (as Schmidt points out) the painful reality of the immense U.S. federal budget deficit with all the consequences that flow from it.

The upshot is that the utilities, faced with both massive regulatory problems and (for the privately owned utilities) the prospective dilution of their shareholders' capital, have de facto been operating on what may be referred to as a *strategy of capital minimization*, i.e., meticulously avoiding commitments that may unnecessarily force them into the capital markets, avoiding commitments that may expose them to criticism and second-guessing by the regulatory bodies.

This is an industry that is now planning on a very slow growth of capacity—and that only in response to experienced growth of demand, rather than to anticipated growth of demand. Such a strategy represents, of course, a radical change from the historic responsibility of the utilities to be ready to satisfy any level of demand of their customers. Some might say the industry can no longer stand prosperity, for prosperity would lead to a further growth of power demand, which would require even more capacity to be financed at even higher real rates of interest.

What then is the prognosis? If we trace the prospective growth of demand and the prospective growth of power supply, we can readily foresee a shortage developing in the early 1990s, possibly even sooner. And that picture would become particularly unsatisfactory if the capacity problem were to occur more or less simultaneously with the prospective tightening of the oil markets. Any such tightening would reinforce the growth in the demand for electric power. That increased demand would, of course, be accelerated by any new oil supply interruption.

This unsatisfactory picture is most notable in Texas and the Southwest, California, and other parts of the country that are growing rapidly. But as a nation we are failing adequately to adjust to that future level of demand. We are not planning ahead. We are not doing now the appropriate advance work on what remain long lead-time items. In the 1990s we are unlikely to have adequate capacity in place. We shall be thrown back into expanding capacity using short-term measures by acquiring more combustion turbines. Combustion turbines will add to the demand for oil, simultaneously imposing the dual problems of rising costs of electric power and insecurity of supply. The final problem regarding the security of supply pertains to the prospective growth in Canadian imports. That growth reflects a sharp change in Canadian policy away from nationalism. But the increased imports do imply a heightened dependence on Canada.

Is this worrisome, as some have suggested—to be so dependent upon Hydro-Quebec? The answer, I believe, is that a modest degree of dependence—a *modest degree* of dependence—is neither risky nor inappropriate. Given the limited capacity of Hydro-Quebec to export power to the Northeast, that dependence is unlikely ever to become very great.

The secretary of energy has regularly warned the country and American utilities about becoming more dependent upon Canadian sources of supply. He has pleaded for more rapid growth of domestic capacity. But both the warnings and the pleadings seem to me largely irrelevant; because of the problems I earlier recounted regarding the regulatory morass and the level of real interest rates, the utilities are quite right to attempt to minimize capital outlays.

The issue that New York State faces is reliability of supply. Some have attempted—regrettably in my judgment—to draw a parallel between the import of electric power from Canada and the import of oil from Middle Eastern producers. There is some degree of superficial parallelism, but it is limited. Canada is a country ultimately characterized by a high degree of stability and predictability in its political arrangements. Moreover, it is close at hand. Canadian supplies are not subject to interdiction as are supplies of oil from the Middle East. Indeed, Canadian imports were specifically excluded from the oil import program, initiated by the Eisenhower administration for national security reasons, because of *unquestioned* security of supply.

To be sure, we have since that time had some experience with Canadian nationalism. One would be ill-advised to assume that the nationalism that resulted in restricted oil exports from Canada throughout the decade of the 1970s has permanently disappeared. In fact, in any future period of supply stringency, it is likely to reappear. But it would not and should not result in a sudden interruption of imports. At most, it would result in a slow decline in Canadian exports. Such restrictions did occur during the 1970s with respect to oil supplies and even, though to a lesser extent, with respect to natural gas supplies. In short, a growing, though measured, dependence upon Canada is relatively free of risk and highly appropriate. Moreover, in the Northeast such dependence is probably unavoidable, given the slow growth of electric power capacity.

It is certainly true that national policy should encourage the creation of greater domestic capacity. But Canadian imports would and should rise anyway because of their low cost. At the moment, because of our complacency, national policy is not moving in that direction. It pays only lip service to the goal of expanding electric generating capacity.

PART II

ELECTRICITY DEMAND, PRICES, AND REGIONAL ECONOMIC ACTIVITY ———————

Part I provides an insightful review of the history of the utility industry and the energy crises as well as of the major forces affecting energy supply and demand in the United States, in general, and of electrical energy in New York State, in particular. We now turn to more detailed analyses of the forces affecting the supply and demand for electrical energy in New York State. Issues associated with the demand for electricity are presented in this section; those associated with the supply of electricity in New York State are covered in Part III.

As in most major industries, planning for the future of the utility industry starts with the industry's forecasts of the demand for the product. These forecasts are the basis for a planning process that usually includes decisions by individual firms about their capital investments and their operating strategies. In general, the higher the capital investment that an industry requires, the more urgent is its need for accurate forecasts of the demand for its products.

One significant reason for this need for accuracy is the relatively long lead time usually associated with making large capital investments productive. Today in New York State it may take as long as eight to twelve years to bring a new steam-powered plant on-line once the decision is made to build the plant. (Reasons for such delays and their various ramifications are discussed in subsequent sections.)

The accuracy of electricity-demand forecasting has been a serious problem in New York State over the past twenty-five years. Because in the 1950s demand growth was underestimated, capacity shortages emerged in the 1960s. Furthermore, before the 1970s the average unit cost of producing electricity generally decreased as new power plants were added to the system and the resulting lower prices stimulated further demand increases. However, since average annual growth rates in demand of 6–8 percent a year were not uncommon through 1970, the societal costs of excessively high demand forecasts were negligible. The forecast demand, if too high for a given time period, usually would be realized one, two, or three years later than originally predicted, and in any case, the new power plant probably would not increase the industry's average unit costs of producing electricity. Thus, little or no penalty was associated with forecasting errors on the high side, whereas significant penalties in terms of the social and economic effects of power outages could be associated with forecasting errors on the low side.

The situation today, unfortunately, is significantly different in New York State and elsewhere. In general, the average unit cost of electricity generated from new power plants is higher than the average unit cost of existing facilities. Thus, the penalty associated with excessively high demand forecasts can be significant (assuming these forecasts are translated into generating capacity with higher than average unit costs).

These relatively new relationships between demand and supply factors have drawn increased attention to the demand-forecasting process. Prior to the mid and late 1970s, New York State consumers or, indeed, the public at large, were generally not interested in the accuracy of forecasts of future electricity demand. Realizing that excess electric-generating capacity would lead to increased rates for all users, consumer groups in New York State (and elsewhere) began to challenge not only the utilities' forecasts but also the methodologies used in generating these forecasts. By engaging specialists from universities and consulting firms, various consumer groups and state agencies were able to provide credible forecasts of electricity demand that were significantly lower than those proposed by the utilities. These lower forecasts were credible because the methodologies used to generate them were, at the time, more sophisticated and comprehensive than those used to generate the utilities' forecasts during that same period.

In the late 1970s, the New York State legislature created the State Energy Office (SEO) and gave it the responsibility for forecasting the overall demand for electricity and other forms of energy in the state. The SEO undertook to develop their own relatively comprehensive models to generate long-range forecasts of the demand for electricity and other forms of energy in New York. The New York Public Service Commission is now required by law to use these biennial forecasts to help establish the need for new electric-generating facilities in the state.

The other major source of forecasts for the long-range demand for electricity in New York State is provided by the New York Power Pool (NYPP), the association of all the publicly-owned and investor-owned electric utilities in the state. The NYPP has also developed their own comprehensive forecasting models.

In Chapter 4, Nancy Meiners provides disaggregated economic and demographic forecasts for New York State to the year 2003. These disaggregated economic and demographic forecasts are exogenous inputs to the NYPP's own comprehensive long-range model for forecasting electricity demand, presented by Nicholas Johnson and John Adams in Chapter 6. Comparable long-range forecasts by the SEO from their own comprehensive models are presented by Charles Guinn in Chapter 5.

Long-range economic forecasts for the United States, which are important exogenous inputs to the SEO model, are obtained from the Wharton Long-Term Forecasting Model. This model also provides similar inputs to the Wharton New York Model (NYMOD), which Meiners uses to generate her forecasts for New York State. The NYMOD model, which provides the long-range economic and demographic forecasts of New York State for the NYPP model, was developed under contract with the NYPP. The use of a common set of long-range economic forecasts for the nation as a basis for the exogenous inputs to both the SEO and NYPP models removes one important possible source of contention should their forecasts not prove comparable. In such cases, discussions can focus on other issues that influence the models' outcomes. However, since there is some variability in the projections and accuracy of individual economic and demographic forecasting models, it might also be useful to have independent checks of these projections too.

In Chapter 7 Timothy Mount and William Deehan provide a brief summary of the recent history of long-range forecasting of electricity demand in New York followed by their own forecasts of the state's electricity demand into the turn of the century. Part II thus offers three sets of independent but comparable forecasts of long-range electricity demand for New York State.

Some interesting observations can be made. First, the models used in generating each of these forecasts are relatively comprehensive and sophisticated, especially in comparison to those previously used by the utilities as the basis for the overly optimistic forecasts prepared fifteen years ago. In contrast to the large differences among the forecasts made in the late 1970s, the long-range forecasts presented by the SEO, the NYPP, and by Mount and Deehan are not all that different from each other.

It is worth noting that the overall structure of the models no longer seems to be the main point of contention, as it was at the end of the last decade. Rather, the main focus of discussion is the relationships among electricity demand, the level of economic activity, and the relative prices of electricity and other sources of energy. That these are difficult issues surrounded by honest disagreements is evident from the chapters in this section. Nevertheless, it is a sign of progress that the issues being discussed are substantive (the relationships among electricity demand, economic activity, and prices) rather than methodological (the structure of the forecasting models).

In the final chapter in this section, Douglas Bohi and Joel Darmstadter provide additional insights into the related issues of the relative prices of electricity versus other forms of energy and the subsequent impacts on demand. These relationships are, of course, important in making predictions about the long-range demand for electricity, though the same relationships will not necessarily hold throughout the country. Another important point addressed by Bohi and Darmstadter in their paper is the long-range supply of oil, upon which a significant portion of the electric-generating capacity of New York State is dependent. Their analysis leads them to somewhat more optimistic conclusions than James Schlesinger arrived at in Part I.

These long-range forecasts of electricity demand and their relationships to prices and economic activity in the state are crucial: they establish a framework for a host of related decisions and actions. These forecasts influence decisions regarding electricity supply, including the type and timing of additional electric-generating capacity. These decisions, in turn, affect environmental and health issues, financial issues, public attitudes toward the industry, and regulatory issues, all of which are treated in detail in subsequent sections.

4

New York State Economic and Demographic Forecasts to 2003

Nancy Meiners

The Wharton New York Model is an annual econometric model covering New York State and northeastern New Jersey. The model was built primarily for long-range forecasting. Four forecasts a year are produced: two ten-year and two twenty-year forecasts. The projections in this chapter are from the twenty-year, September 1984, Wharton New York Model forecast.

There are thirteen model regions. Eight are Metropolitan Statistical Areas (MSAs) or combinations of MSAs.* The seven New York State MSAs modeled are the Nassau/Suffolk MSA (Long Island), the five boroughs of New York City, the Albany/Schenectady/Troy MSA, the Syracuse and Utica/Rome combined MSAs, the Rochester MSA, the Buffalo MSA, and the combined Binghamton and Elmira MSAs. The non-MSA portion of New York State is divided into regions which, with the addition of the closest MSA, form utility service territories. For example, the Buffalo MSA is one model region; the rest of Niagara Mohawk's western service territory is another. The model forecasts employment at the 1-digit SIC code level for all regions. For the MSAs, employment for the major 2-digit manufacturing sectors is forecast. In addition, demographic variables, components of personal income, gross regional product, consumer price index, households, and household employment are projected.

Employment is explained by a set of national variables, by regional total demand, and by a regional cost-of-doing-business index. This index

---

*The revised metropolitan definitions of the U.S. Office of Management and Budget went into effect on January 1, 1985. Accordingly, the previous "Standard Metropolitan Statistical Area," or "SMSA," is now referred to as the "Metropolitan Statistical Area," or "MSA."

includes measures of capital costs, labor costs, and energy costs. Energy prices are an input to the forecast. The energy price index is a weighted average of costs of electricity, coal, natural gas, #2 and #6 fuel oils, liquified petroleum gas, and gasoline. The weights vary by sector.

This Wharton New York Model forecast is based on the July 1984 Wharton Long-Term Model Outlook for the United States.

The New York State economy is expected to perform better during the upcoming ten years than over the last decade, primarily because the U.S. economy is expected to be stronger. U.S. economic growth, measured by the growth in real gross national product, is projected to average 2.9 percent a year during the upcoming decade, versus 2.0 percent over the previous decade. Real GNP is expected to grow by 6.3 percent in 1984 and by 2.7 percent in 1985. (See Table 4.1)

There are several reasons for the stronger U.S. growth expected for the next decade.

- Real defense spending growth is expected to be much more robust—4.0 percent a year, versus 2.1 percent during the previous decade.
- Despite cuts in nondefense spending programs, total federal spending grows more rapidly—3.6 percent a year, versus 2.1 percent.
- Effective personal tax rates have been cut by 23 percent, producing a consumer spending boom.
- Liberalized depreciation rules have reduced the cost of business capital, and investment activity is very strong.
- While the Federal Reserve Board continues to pursue a relatively tight monetary policy, current and projected low inflation rates imply that interest rates will increase only moderately.
- While the risk of disruption to world oil supplies due to a war in the Middle East is substantial, no sharp increases in world energy prices are built into the baseline U.S. forecast.
- Despite the assumptions that the U.S. Senate and House will agree in 1985 on a fiscal policy package that will reduce the deficit by $160 billion, and that "permanent" surtaxes of $30 billion will be enacted in 1986, fiscal policy remains stimulative.
- The strong economy provides substantial increases in state and local government revenues, and these revenues allow these governments to expand their spending programs—2.5 percent a year, versus 1.1 percent.

This strong growth is limited in the near term by a strong upsurge in imports, but devaluation of the dollar over the next several years will reduce imports and increase exports.

U.S. economic performance during the upcoming decade could be significantly less than projected in the baseline forecast, if

TABLE 4.1 Output and Employment Forecast for New York State (Average Annual Percent Change)

	Output			Employment		
	1974–83	1984–93	1994–2003	1974–83	1984–93	1994–2003
United States	2.0	2.9	2.4	1.8	1.8	1.1
New York	0.9	2.3	2.8	0.2	1.2	1.2
Downstate New York	0.9	2.1	2.7	0.1	1.0	1.1
Upstate New York	0.8	2.9	2.9	0.4	1.4	1.2

- oil supplies from the Middle East are disrupted by a major war;
- the Federal Reserve Board pursues too stringent a monetary policy, resulting in very high interest rates;
- the U.S. Congress and the president place top priority on reducing the U.S. deficit and raise taxes sharply or cut spending significantly;
- management agrees to excessive wage rate settlements with the unions to avoid strikes in the face of strong demand, thereby setting in motion another inflationary spiral.

There are also upside risks associated with the forecast—such as the other developed countries adopting strong domestic growth policies, leading to stronger U.S. export performance—but most of the plausible risks seem to be on the downside.

Real U.S. growth slows slightly to 2.4 percent a year during 1994–2003, but employment growth in New York State continues at a rate almost equal to its 1984–93 pace. The continued strength in New York State is concentrated in the downstate region with its large service and financial sector component. Upstate New York growth slows slightly following the U.S. pattern.

New York's improved performance can be attributed in part to a reduction in its cost disadvantage relative to the rest of the country. New York is not a low-labor-cost area and will not attract industries needing a large unskilled or semiskilled work force. Its cost disadvantage versus the West, Southwest, and Southeast, however, has been reduced or eliminated. Living costs in California substantially exceed those in New York. As a result, the New York economy will lose fewer jobs to these other regions and will actually gain employment from other areas.

The relatively weak performance of the New York economy in the previous decade is due in part to the closings of older, uncompetitive plants such as the steel mills in Buffalo. Most of these facilities have been closed or have been cut back to fractions of their former size. While some more jobs will be lost, the worst appears to be over. In addition, New York City no longer has the financial problems it experienced in the past decade. The outmigration from the state to the Sunbelt is expected to slow in the next decade with positive net migration into the state showing by 1992.

A lot of New York businesses that demand unskilled, low-wage labor have already relocated to regions where wages are lower. Many companies that remain and those attracted to the state depend on the pool of highly skilled labor which is New York's greatest asset for growth. The state economy includes strong firms in growth industries. Many of these firms are expanding in New York or have indicated intentions to remain in the state. Examples include Kodak in Rochester, IBM in the Lower

Hudson Valley and elsewhere, and GE in Albany/Schenectady/Troy and elsewhere.

The state also benefits from the stronger defense spending anticipated for the upcoming decade. A stronger U.S. defense posture has been endorsed by both parties, with Walter Mondale calling for 4 percent real growth and Ronald Reagan calling for 5 percent real growth. Major beneficiaries of this high defense spending include Grumman in Nassau/Suffolk, GE in Albany/Schenectady/Troy, and other high-technology firms.

All the major industrial sectors in New York State exhibit improved or continued strong performance during 1984–93 versus 1974–83. The relative growth in employment by sector is shown in Table 4.2.

Although manufacturing employment in the state declines during the upcoming decade, the rate of decline is much reduced from the previous decade. Nonmanufacturing employment growth rises from 1.0 percent a year to 1.8 percent during the upcoming decade. All components of nonmanufacturing show improved growth except services, which still remain strong. Services and finance, insurance, and real estate exhibit strong growth over the next decade, continuing the trend established over the last ten years. Wholesale and retail trade rebounds from a mediocre to a very good performance. Employment in contract construction and in transportation, communications, and utilities declined during the last decade. Employment in construction grows significantly after the severely depressed employment levels of 1984. Employment growth in transportation, communications, and utilities is expected to be concentrated almost entirely in communications.

Overall, private nonfarm employment growth improves to 1.7 percent a year during the 1994–2003 period. This improvement results from

TABLE 4.2 New York State Employment, by Sector (Average Annual Percent Change)

	1974–83	1984–93	1994–2003
Private Nonfarm	0.3	1.1	1.7
Manufacturing	−2.1	−0.3	−0.8
Nonmanufacturing	1.0	1.8	1.8
Contract construction	−2.3	2.6	2.1
Transportation, communications, utilities	−1.4	0.1	−0.3
Wholesale and retail trade	0.3	1.6	1.7
Finance, insurance, real estate	1.6	1.8	2.5
Services	2.7	2.2	1.9

continued strong growth in its nonmanufacturing component. Manufacturing employment continues to slip downward, but the nonmanufacturing component dominates the total.

Historically, population growth in New York has been weak. In the past ten years population growth has been slightly positive only in 1981. This negative historical trend is expected to reverse itself over the next twenty years, with a net projected population gain of over 1.2 million by 2003.

On the substate level the regions with the most growth potential are located around New York City. Long Island has been experiencing a growth spurt recently, but this region is outshone by the area northwest of the city. In the Putnam/Orange/Rockland/Westchester combined county region, employment is projected to grow about 2 percent a year over the next twenty years. In the Dutchess/Ulster county area, expected annual employment growth is 2.7 percent with 4.3 percent of that average total in nonmanufacturing employment. The location of the Dutchess/Ulster region has many advantages. It is between Albany and New York City, on the New York State Thruway, and on water. There is already a lot of IBM activity in the region. Dutchess/Ulster is probably on its way to becoming a mini-Silicon Valley.

The regions projected to have very low employment growth are Buffalo and New York City. Buffalo has had a lot of problems in recent years, mainly centered in its primary metals industry. It lost 6,000 jobs in 1982 alone when Republic Steel closed. Much of the remaining metals activity is in fabricated metals—foundries supplying the auto manufacturing companies that have remained in the region. Their future is very closely tied to the continued health of the domestic auto market. With this one area of uncertainty aside, Buffalo is not expected to experience any more major job losses like those suffered during the previous decade. The Buffalo economy will begin to rebuild. This process may be slow as a result of caution on the part of companies scouting new locations: uncertainties about government fiscal soundness and stability, rising crime, and so on. Once they are satisfied on these points, companies will be attracted by Buffalo's locational advantages and, just as important, by the current availability of buildings; land, water, and sewer capacity; and labor freed up by the decline of the local steel industry.

For many years manufacturing employment in New York City has been declining as a result of land prices, the effects of congestion on transportation costs, and so on. The city has a greater concentration of its total employment in the services sector than does the state (one-third compared to one-quarter), and a larger financial sector also. Employment in these sectors will continue to grow but at slower rates than in recent history. Population is projected to remain fairly constant in the next two decades, with the city gaining only 13,000 people over that period.

New York State has a pool of highly skilled labor, a large number of universities to train new entrants for this pool, and many amenities to keep this pool attracted to the state. Probably the greatest potential for state employment lies in attracting the types of businesses needing skilled labor. This comparative advantage of the state should fuel the strong growth rates expected during the next two decades.

5

New York Electricity and Energy Forecasts —————

Charles R. Guinn

Planning for long-range electricity supply and evaluation of specific electricity generation or transmission projects both require a long-range electricity demand forecast with sectoral, regional, and temporal detail.* Electricity demand forecasts must be sensitive to changes in patterns of economic activity by sector, changes in demographic patterns, changes in electricity and fossil fuel prices, changes in end-use efficiencies, and new end-use technologies. Moreover, electricity demand forecasts, to be useful for planning and project evaluation, must provide substate detail at the utility service area level and sufficient temporal detail to determine load curves.

Electricity demand forecasts must be the product of a comprehensive and integrated forecasting system that examines the interrelationships among energy use prices, fuel choice, economic activity, and public policy. It is especially important for public policy analysis that the forecasting methodology explicitly examine the impact of mandatory end-use regulation and expected end-use efficiency improvements.

New York State has developed and refined an independent energy demand forecasting capability and expertise as a result of the State Energy Master Plan (SEMP) process. The State Energy Office (SEO) has prepared a very detailed set of biennial energy forecasts since 1979. A new forecast is being developed for 1985. The new forecast, like previous forecasts, will examine electricity demand within the framework of future economic and demographic patterns, total energy demand, future energy prices, end-use efficiency, and conservation programs.

---

*See the end of the chapter for relevant tables and figures.

The New York State Energy Forecasting System, developed by the SEO over the past six years, consists of a series of models and data bases that must be updated on an ongoing basis. The system includes separate energy demand models for each of the four major energy consuming sectors—residential, commercial, industrial, and transportation, as well as an electricity-load-curve forecasting model and an underlying state macroeconometric model that provide forecasts of the basic measures of economic activity that drive, in part, the various energy demand models. The Long Run Annual Economic Forecasting Model developed by Wharton Econometric Forecasting Associates supplies the national economic activity inputs for the state macroeconomic model.

The residential model provides a unique framework for forecasting residential energy consumption by eleven end-uses, four building types, and five fuel types for each electricity service territory in the state (see Table 5.1). The model consists of three submodels: space heating, water heating, and appliances. The model relates energy requirements by end-use to the following factors: forecasts of customer growth, an economic assessment of fuel choice and appliance saturations, and forecasts of unit energy consumption based on prospective efficiency improvements and mandatory end-use regulations.

Specific residential end-uses include: devices for space heating, water heating, and air conditioning; refrigerators; freezers; ranges; clothes dryers; dishwashers; televisions; lighting; and a residential category including other miscellaneous uses of electricity.

The commercial sector model provides a framework for forecasting commercial energy consumption by eight end-uses and three fuel types for each electric utility service territory in the state (see Table 5.2). Eight building types are analyzed: private office; retail/wholesale; health care; educational; state, federal, and local government; and a final category of miscellaneous buildings. The model relates energy consumption by end-use to the following factors: a forecast of commercial sector floorspace additions, an economic assessment of space and water-heating fuel choices, projections of other end-use saturations, base year unit energy consumption, and annual changes in the base-year unit energy consumption resulting from changes in utilization, efficiency improvements, and mandatory end-use regulations.

The commercial end-uses include: devices for space heating, cooling, ventilation, and water heating; ranges; refrigerators; lighting; and a miscellaneous category including such auxiliary uses as data processing.

The industrial model employs an econometric approach to forecasting electricity consumption by electricity service territory and natural gas and oil consumption on a statewide basis. This approach relates industrial energy consumption to forecasts of economic activity (output) and relative fuel prices.

An overview of the 1983 SEMP energy demand forecast, which is currently being updated for 1985, provides some perspective on future electricity demand (see Table 5.3). Total energy demand or consumption is forecast to decline by an average rate of 0.3 percent a year from 1982 to 1999 (see Table 5.5). Electricity consumption (sales) is forecast to increase at an average rate of 1.3 percent a year over the forecast period, while total statewide electricity peak demand is projected to increase at an average rate of 0.9 percent a year. The electricity demand forecast for specific end-uses varies considerably and ranges from an increase of 4.7 percent a year for residential space heating to a decrease of 0.4 percent a year for freezers (see Table 5.4).

The SEMP electricity demand forecasts are used for many purposes and applications in addition to the supply planning activities associated with the SEMP process. Under Article 8 of the New York Public Service Law, the SEMP electricity demand forecasts are binding on the State Board on Electric Generation Siting and the Environment with respect to determination of need for future steam-powered electric generating facilities. Also, the projected electricity and natural gas demand forecasts are binding on the Public Service Commission with respect to any determination of need for major electric and gas transmission facilities under Article 7 of the Public Service Law. In addition, however, the SEMP demand forecasts have been used in a variety of federal and state regulatory proceedings such as the Federal Energy Regulatory Commission (FERC) Boundary Gas Proceedings, the FERC Prattsville Proceedings, and the Public Service Commission Niagara Mohawk Long-Term Avoided Costs Proceedings. Moreover, the forecasts have been used in a number of specific policy studies such as the Nine Mile Point Two Economic Study conducted by the SEO, the Marburger Panel on Shoreham, and the Millonzi Commission on hydro-power reallocation.

TABLE 5.1 Detail in Residential Energy Demand Model

End-Use	Fuel Type	Building Type
Space heating	Electricity	Single Family
Water heating	Natural gas	2–4 Units
Air conditioning	Oil	5–49 Units
Refrigerator	Wood	50+ Units
Freezer	Solar	
Range		
Clothes drying		
Dishwasher		
Television		
Lighting		
Other		

TABLE 5.2 End-Use/Fuel/Building Detail in the SEO Commercial Model

End-Use	Fuel Type	Building Type
Space heating	Electricity	Office
Cooling	Natural gas	Retail/wholesale
Ventilation	Oil	Healthcare
Water heating		Education
Range		State government
Refrigerator		Local government
Lighting		Federal government
Other		Miscellaneous

**TABLE 5.3   Forecast of New York State
End-Use Energy Consumption in 1999 by
Sector and Fuel Type (Trillions of BTUs)**

Consumption By Sector and End-Use	Electricity	Natural Gas	Petroleum Products	Other[a]	Total
Residential[a]	140.2	298.3	198.9	58.3	695.7
Space heating	15.4	192.0	144.3	58.2	409.9
Water heating	14.1	79.1	54.6	0.1	147.9
Range	4.9	22.8	–	–	27.7
Clothes drying	8.1	4.4	–	–	12.5
Air conditioning	11.9	–	–	–	11.9
Refrigerator	26.8	–	–	–	26.8
Dishwasher	3.5	–	–	–	3.5
Freezer	7.2	–	–	–	7.2
Television	10.5	–	–	–	10.5
Lighting	14.3	–	–	–	14.3
Other	17.7	–	–	–	17.7
Commercial[b]	171.5	126.5	100.0	11.4	409.4
Space heating	17.6	103.5	90.4	11.4	222.9
Air conditioning	41.5	–	–	–	41.5
Water heating	3.3	9.3	9.6	–	22.2
Lighting	61.1	–	–	–	61.1
Other	32.2	13.7	–	–	45.9
Industrial	123.5	95.6	145.5	100.4	465.1
Transportation	15.0	–	887.6	–	902.6
Highway	2.3	–	571.3	–	573.6
Auto	1.5	–	362.4	–	363.9
Truck	.8	–	208.9	–	209.7
Air	–	–	188.9		188.9
Vessel	–	–	53.7	–	53.7
Intercity rail	–	–	9.7	–	9.7
Transit	8.4	–	14.8	–	23.2
Other	4.3	–	49.2	–	53.5
Other[c]	4.9	–	–	–	4.9
Total	455.2	520.4	1,332.0	170.1	2,477.7

[a]Includes wood, solar, and coal.

[b]NYPA is included in residential and commercial total electricity consumption but not in end-use consumption. Load management impacts are included in residential and commercial total electricity consumption but not in end-use consumption.

[c]Other includes state and highway lighting and sales for resale.

TABLE 5.4 New York State Electricity Consumption (Sales) by Sector[a] and End-Use, 1982–1999 (Trillions of BTUs)

Sector of Energy Use	1982	1987	1991	1995	1999	Growth Rate (%) 1982–99
Residential	113.8	117.7	124.8	133.3	140.2	1.2
Space heating	7.1	8.9	11.0	13.6	15.4	4.7
Water heating	8.2	8.8	10.2	12.5	14.1	3.2
Range	4.4	4.5	4.5	4.7	4.9	0.6
Clothes drying	6.2	6.9	7.4	7.8	8.1	1.6
Air conditioning	8.9	9.4	10.3	11.1	11.9	1.7
Refrigerator	28.2	27.4	26.8	26.3	26.8	-0.3
Dishwasher	2.4	2.6	2.8	3.2	3.5	2.2
Freezer	7.7	7.5	7.4	7.3	7.2	-0.4
Television	9.2	9.0	9.1	9.8	10.5	0.8
Lighting	13.3	13.3	13.9	14.3	14.3	0.4
Other	14.7	15.4	16.2	17.1	17.7	1.1
Commercial	145.5	149.7	156.2	162.9	171.5	1.0
Space heating	12.5	13.1	13.9	15.3	17.6	2.0
Air conditioning	31.3	33.9	36.7	39.2	41.5	1.7
Water heating	2.2	2.2	2.3	2.6	3.3	2.4

TABLE 5.4 (*Continued*)

Sector of Energy Use	1982	1987	1991	1995	1999	Growth Rate (%) 1982–99
Lighting	50.8	53.1	55.1	57.8	61.1	1.1
Other	32.0	30.4	30.6	31.2	32.2	0.0
Industrial	90.2	99.3	108.2	118.1	123.6	1.9
Transportation	9.7	10.9	12.3	13.9	15.0	2.6
Other[b]	4.6	4.2	4.7	4.8	4.9	0.4
Total electricity Consumption (Sales)[c]	365.4	381.8	406.2	433.0	455.2	1.3

[a]This is an estimate of electricity sales (including the impact of load management) for the residential, commercial, industrial, and transportation sector on a statewide basis. The sector end-use data do not include the impact of load management. Therefore, the sum of the end-use data does not equal the sector estimate.

[b]Other includes street and highway lighting and sales for resale.

[c]Total includes Jamestown and Freeport.

TABLE 5.5 New York State Energy Consumption and Economic Activity, 1965–1999: Selected Measures

Selected Measure	Actual Data (Average Annual Percent Change)				Projected
	1965	1973	1978	1982	1999
Total primary energy consumption (Trillions of BTUs)	3,539.0	4,281.5 (2.4)	4,101.8 (−0.9)	3,553.5 (−3.5)	3,553.3 (0.0)
End-use energy consumption (Trillions of BTUs)	2,946.9	3,400.7 (1.8)	3,190.3 (−1.3)	2,615.1 (−4.8)	2,477.7 (−0.3)
Electricity consumption (Sales) (Trillions of BTUs)	224.0	341.2 (5.4)	357.2 (0.9)	365.4 (0.5)	455.2 (1.3)
Gross state product (GSP) (Millions of 1982$)	220,770	270,086 (2.6)	265,326 (−0.4)	274,826 (0.9)	393,997 (2.1)
Personal income (Millions of 1982$)	156,543	200,884 (3.2)	204,474 (0.4)	214,228 (1.2)	281,681 (1.6)
Per capita income (1982$)	8,798	11,041 (2.9)	11,538 (0.9)	12,131 (1.3)	16,438 (1.8)
Ratio of primary energy consumption to GSP (10 BTUs/1982$)	16.0	15.9 (−0.1)	15.5 (−0.5)	12.9 (−4.9)	9.0 (−2.1)

TABLE 5.6 Forecast of New York State End-Use Energy Consumption by Sector, 1982–1999

Sector of Energy Use	Trillions of BTUs		Average Annual Percent Change 1982–1999
	1982	*1999*	
Residential	809.2	695.7	−0.9
Commercial	429.4	409.4	−0.3
Industrial	367.8	465.1	1.4
Transportation	1,002.5	902.6	−0.6
Other[a]	4.6	4.9	0.4
Total end-use energy consumption[b]	2,615.1	2,477.7	−0.3

[a]Other includes street and highway lighting and sales for resale.
[b]Total includes Jamestown and Freeport.

TABLE 5.7 New York State End-Use Energy Consumption by Fuel Type, 1982–1999

Fuel Type	Trillions of BTUs		Average Annual Percent Change 1982–1999
	1982	*1999*	
Electricity	365.4	455.2	1.3
Natural gas	614.1	520.4	−1.0
Petroleum products	1,512.4	1,332.0	−0.7
residential, commercial, and industrial	519.6	444.4	−0.9
transportation	992.8	887.6	−0.7
Coal/solar	51.1	66.0	−1.5
Total end-use energy consumption	2,615.1	2,477.7	−0.3

TABLE 5.8 New York State Electricity Peak Demand and Growth Rates by Utility, 1982–1999

Utility	Summer Peak (MW)			Winter Peak (MW)			
	1982	1999	Growth Rate (%)	1982	1999	Growth Rate (%)	Peak Growth (%)
Central Hudson Gas & Electric	666	933	1.3	630	883	2.0	1.7
Consolidated Edison	7,326	7,044	-0.2	4,920	4,895	0.1	-0.2
Long Island Lighting	3,045	3,646	1.1	2,471	2,921	1.0	1.1
New York State Electric & Gas	1,771	2,645	2.4	2,090	3,396	2.9	2.9
Niagara Mohawk	4,708	5,513	0.9	5,223	6,493	1.3	1.3
Orange & Rockland	712	850	1.0	509	662	1.6	1.1
New York Power Authority	2,377	3,108	1.6	2,488	3,410	1.9	1.9
New York Power Pool coincident peak[a]	21,252	24,543	0.9	19,320	23,862	1.3	0.9

[a]Includes village of Freeport and city of Jamestown; these loads are included in the NYPA forecast beginning in September 1988.

TABLE 5.9   New York State Annual Electricity Consumption (Sales) by Utility, 1982–1999

Utility	1982	1999	Average Annual Percent Change
Central Hudson Electric & Gas	3,390[a]	4,824	2.1
Consolidated Edison	27,340	28,425	0.2
Long Island Lighting	12,519	15,285	1.2
New York State Electric & Gas	10,711	17,102	2.8
Niagara Mohawk	28,082	35,180	1.3
Orange & Rockland	2,902	3,818	1.6
Rochester Gas & Electric	5,238	7,650	2.3
New York Power Authority (NYPA)[b]	16,446	21,129	1.3
New York Power Pool[c]	107,108	133,413	1.3

[a]Gigawatt-hours.
[b]NYPA includes Jamestown and Freeport beginning September 1988.
[c]Includes Jamestown and Freeport.

FIGURE 5.1   Energy Flow for New York State, 1983 (Trillions of BTUs)

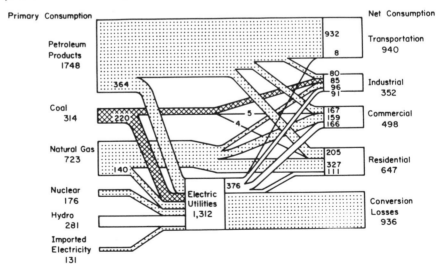

Source: Prepared by the author.

FIGURE 5.2 End-Use by Fuel for New York State, 1983 (Trillions of BTUs)

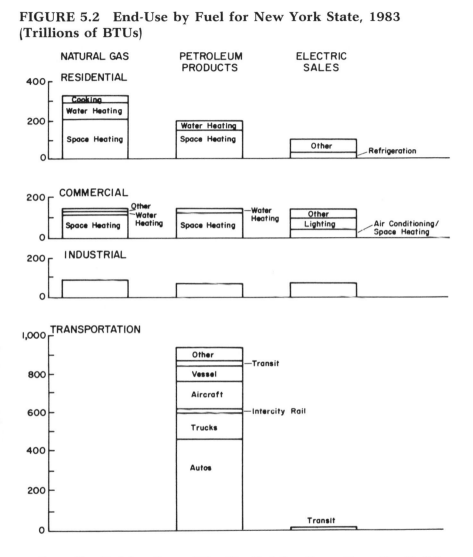

Source: New York State Energy Office, New York State Energy Master Plan, Draft Report, August 1983.

FIGURE 5.3   New York State Energy Forecast

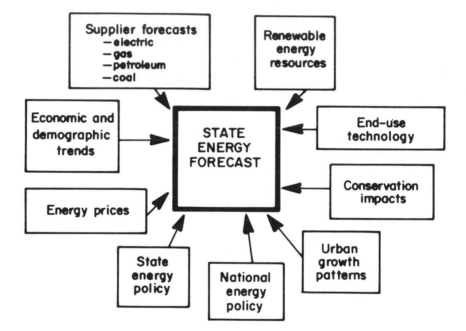

Source: Prepared by the author.

FIGURE 5.4 Wharton Long-Term Forecast

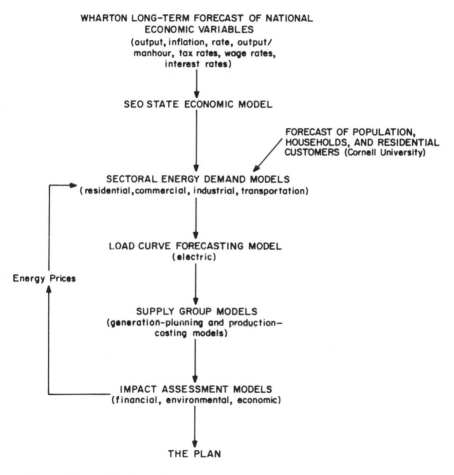

Source: Prepared by the author.

FIGURE 5.5 Energy Forecasting for New York State

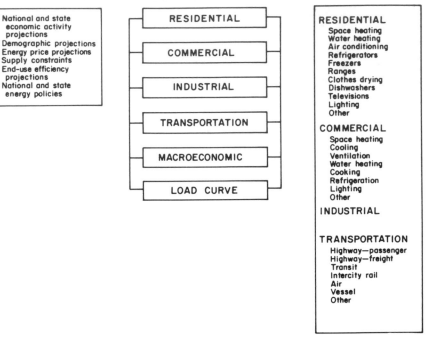

ASSUMPTIONS

National and state
  economic activity
  projections
Demographic projections
Energy price projections
Supply constraints
End-use efficiency
  projections
National and state
  energy policies

FORECASTING MODELS

RESIDENTIAL

COMMERCIAL

INDUSTRIAL

TRANSPORTATION

MACROECONOMIC

LOAD CURVE

END-USE DETAIL

RESIDENTIAL
  Space heating
  Water heating
  Air conditioning
  Refrigerators
  Freezers
  Ranges
  Clothes drying
  Dishwashers
  Televisions
  Lighting
  Other

COMMERCIAL
  Space heating
  Cooling
  Ventilation
  Water heating
  Cooking
  Refrigeration
  Lighting
  Other

INDUSTRIAL

TRANSPORTATION
  Highway—passenger
  Highway—freight
  Transit
  Intercity rail
  Air
  Vessel
  Other

Source: Prepared by the author.

FIGURE 5.6 Electricity Sales by Sector, 1983

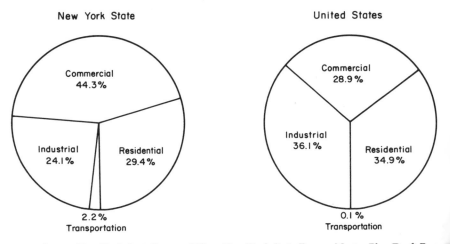

New York State

Commercial
44.3%

Industrial
24.1%

Residential
29.4%

2.2%
Transportation

United States

Commercial
28.9%

Industrial
36.1%

Residential
34.9%

0.1%
Transportation

Source: New York State Energy Office. New York State Energy Master Plan Draft Repo
August 1983.

6

Electricity Demand Growth in New York State: The Uncertain Factor in the Electricity Planning Process ——————

Nicholas Johnson
John Adams

Today's forecasters of electricity demand growth are faced with the formidable task of developing reasonable forecasts for their corporations without knowing what the future will bring. As a consequence, there is some certainty about the future: it will differ from what the forecaster predicts. It is this factor that forecasters and planners must consider when developing long-term plans.

Figure 6.1 illustrates very clearly the difficulties and uncertainties of forecasting in turbulent times. This figure displays the actual kWh sendout for the member systems of the New York Power Pool (NYPP) for the years 1983 and (year-end estimated) 1984, the forecast made by the NYPP in 1982 and updated in 1983, and the forecast made by the New York State Energy Office (SEO). These forecasts currently form the basis for the long-range electric planning process in New York. As can be seen, the estimated 1984 sendout has already exceeded the NYPP's forecast for 1987 and has equaled the SEO's forecast for 1989.

These forecasts were made when the U.S. economy had begun to recover in 1982 from the deepest recession since the Great Depression. The consensus of the experts was that any recovery would be weak at best and very short-lived as a result of high real interest rates. Well, a funny thing happened. Not only did the U.S. economy experience a strong post-World War II-type recovery, but New York's share of U.S. economic growth rebounded as well.

Figure 6.1 also illustrates another concern with the forecasting process, namely, the tendency to become hypnotized by current events. This tendency to minimize growth when one's current view of the world is pessimistic, or vice versa, provides all the more reason to recognize the uncertainty associated with the future. Too often a forecast is used to

FIGURE 6.1 NYPP and SEMP Projections, 1983–1993, and Actual kWh Sendout 1983 and 1984 versus Time

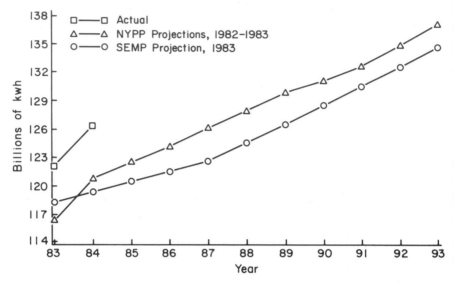

pave the future road with the past rather than to stay open to future possibilities.

The growth in the demand for electricity is tied very closely to the nature and growth of economic activity. This relationship continues to change. The economy now requires less electricity to produce one unit of output. However, the linkage between the two remains strong even as the economy becomes more efficient in the use of its energy inputs.

The implicit ratio of electricity growth to economic growth in both the U.S. and New York State has declined from a level greater than 1.0 to a current level of approximately 1.0 and is projected to decline in New York State to a level of less than 0.9 by the end of the forecast period. A recent survey by the Edison Electric Institute of several national long-term electricity demand forecasts found implicit ratios between the growth in electricity demand and economic growth ranging between 0.8 and 1.91, with the most frequently occurring value around 1.0.

The link between economic growth and the growth in the demand for electricity is an issue that continues to be strongly debated. In fact, the most recently published NYPP and SEO electricity demand forecasts project the long-term growth rates for kWhs to be 1.4 percent and 1.3 percent respectively. The small difference in growth rates notwithstanding, the two forecasts are developed from two entirely different economic scenarios. The NYPP's view of the future is much more pessimistic than the SEO's. This means that the implicit ratio of electricity growth to economic growth contained in the SEO forecast is considerably less than 1.0. Stated another way, the SEO forecast assumes greater energy efficiency gains will be achieved over the planning period. In fact, the implicit ratios in the current SEO forecast are 0.47 for the period 1982–87, 0.68 for the period 1987–91, and 0.69 thereafter.

This divergence of views on the potential efficiency gains evolves from the values and judgments made by the forecasters and their sponsoring institutions when comparing gains that are theoretically feasible to those that may be economically feasible and to what is eventually made available in the marketplace. This type of new product availability involves such issues as market efficiency and market imperfections. The difference of opinion that exists between the industry and the SEO explicitly embodied in the forecasts. However, it should be noted that the ratios implicit in the SEO forecast are outside the low extremes of the Edison Electric Institute survey. In fact, the turnaround in the growth of productivity now being projected by almost all economists implies a stabilization in this trend, if not an outright reversal. This projection is further reinforced by a study by Dale W. Jorgenson of the Department of Economics at Harvard University that was published in the July 1984 issue of Energy Journal. The study reports "our first and most important conclusion is that *electrification* plays a very important role in productivity growth" (emphasis added).

The final point we would like to make regarding the electricity/economy interaction is that the economy of New York State consists of several diverse regional economies. Thus, the customer mix and load characteristics of the NYPP utilities serving these regional economies will vary depending on the nature of the economic activity of the region and on other environmental factors. For instance, the upstate regions of New York, in comparison to downstate regions, can be characterized accordingly: regional economies more oriented toward manufacturing; differing mix of urban, suburban, and rural populations; lower incomes; lower electric prices; and lower cooling degree-day requirements. Thus, the upstate utility systems tend to peak in winter, and have more balanced summer and winter loads and, consequently, higher annual load factors. The downstate utility systems peak more strongly in summer and have relatively low annual load factors. Finally, the upstate economies tend to be more cyclical than downstate economies.

Figures 6.2 and 6.3 reinforce and illustrate graphically the historical relationship between economic growth and the growth in the demand for electricity in the post-oil-embargo era. In Fig. 6.2 we plot electrical energy in kWh and gross state product (GSP) versus time for the decade 1974–

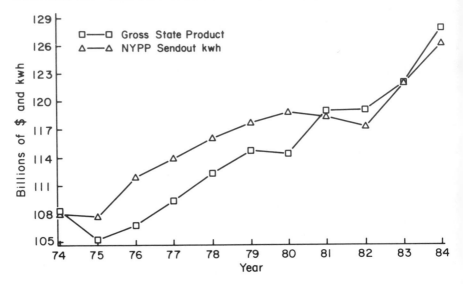

FIGURE 6.2 GSP and NYPP Sendout versus Time

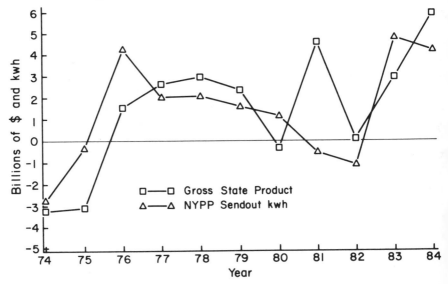

FIGURE 6.3. GSP and NYPP Sendout:
Annual Changes versus Time (1974–1984)

84. Although GSP and kWh do not move in lockstep on a year-to-year basis, the long-term relationship over the last decade certainly has been very close to 1:1. In fact, over the last decade GSP has expanded at an annual compounded growth rate of 1.7 percent, while electricity demand, measured in terms of kWh, has expanded at an annual compounded growth rate of 1.6 percent. Given the measurement error involved, the two have grown statistically on a 1:1 basis over the last decade.

Figure 6.3 represents the year-to-year change. Again, there is consistent movement between GSP and kWh. In general, when economic activity is rising, the demand for electricity is rising, and when economic activity declines or falters, the demand for electricity declines or falters. This is true except for the period 1980–81, which followed the "oil shock" and resultant recession of 1979. The period was preceded by a significant increase in energy prices and characterized by a service-sector-based recovery that was weak and short-lived. Also, manufacturing employment in New York State had declined precipitously prior to this time, and continued to decline during and beyond this time.

Table 6.1 presents historical and projected growth rates for key economic and demographic parameters for the decades preceding and following 1984. These estimates are based on the September 1984 forecast of the New York State Economic Model prepared by Wharton Econometric Forecasting Associates. The figure also presents the historical growth rate of electricity demand and the rate at which demand would have to grow from the 1984 year-end estimate to reach the level currently being forecast by the NYPP and the SEO in 1994. As is indicated, the economic outlook for New York State for the next decade has improved considerably over the previous decade. If one looks at the nine-year period prior to 1983, a period characterized by two major energy price shocks and three recessions, it is clear that the outlooks for the U.S. and New York economies are significantly improved. The question then becomes, how fast will electricity demand grow given this improved outlook?

Before answering this question, we should identify the uncertainties that may affect the economy in the future. These include, among others:

- Cumulative impact of U.S. deficits
- Tax policy (federal and state)
- "Debt overhang"
- Interest rates (monetary policy)
- Acid rain legislation
- Energy prices
- Inflation
- World economic order (protectionism).

**TABLE 6.1 Key Economic Assumptions:
Historical and Projected**

Factors	Average Annual Growth Rates (%)	
	1974–84	1984–94
Economic:		
GNP	2.8	2.8
NY GSP	1.7	2.2
NY Employment	0.7	1.0
U.S. Labor Productivity	0.8	1.3
Real Electricity Price	3.7	0.3
Inflation	9.0	5.0
Demographic:		
Population	−0.2	0.2
Households	N/A[a]	0.5
Electricity Demand:		
NY GWh	1.6	
Projected:[b]		
NYPP		1.0
SEMP		0.8

[a]Not available.
[b]Based on 1984 year-end estimates and most current forecast.

These factors, which are not necessarily mutually exclusive, could quickly undermine our improved outlook, evoking a replay of the period from the mid-1970s to early 1980s. Many of these factors have been discussed and debated at length in the economic literature. Their potential is real but their ultimate outcome and impact are highly uncertain.

What does the NYPP see as the outlook for the growth in electricity demand for the coming decades? If the economy of New York State can sustain net real growth in the coming decades, the demand for electricity will expand significantly faster than it did in the 1970s. This conclusion is consistent with a recent study by Data Resources, Inc., entitled *Analysis of Historical Electric Trends* and published in November 1983. The study found that since 1972 the slowdown in the growth in the demand for electricity can be attributed primarily to the poor performance of the U.S. economy. Thus, a robust and expanding economy will require increasing electricity output. As a result, the NYPP's most likely scenario shows moderate growth caused by: economic expansion; continued growth in the number of households; and continued substitution of electricity for oil and gas. The exact growth path for electricity demand, on the other hand, will be highly uncertain and will be the result of countervailing forces.

The sources of uncertainty affecting the determination of a future growth path are many. Some of them have already been mentioned. The others include:

- Random events: these would include such factors as weather, natural disasters, and regional wars.
- Technological innovation: this is a key unknown. It is the material that fuels robust economic growth. Also, from the point of view of electricity consumption, the NYPP sees the introduction of electricity-consuming technological innovations as counterbalancing some of the efficiency gains that will be achieved. This counterbalancing will occur within all sectors of the economy. For instance, the ever-present electronic keyboard, heating and melting of materials, space and water conditioning, new processes, and so on.
- Forecast models: the models used to project future electricity demand, no matter how sophisticated, are strongly rooted in historical data and relationships. Thus, the structure that is appropriate for the period from which the model is developed will not be totally adequate for projecting future demand. That is, structural change has occurred in the past and will occur in the future. Also, the models are statistical in nature and, therefore, subject to statistical errors in model parameters and measurement errors in input data.

Given that uncertainty exists, we need to analzye it to get a reasonable range for the growth paths that the demand for electricity could follow in the coming decades. Obtaining such a range is important because small changes in growth can have a large effect. Finally, all that can be said about the future with any certainty is that it will be different from the past. Technological change is occurring at a pace unparalleled in history and the direction that change will take is highly debatable.

How do we analyze uncertainty? One approach that we would like to propose is to develop alternative scenarios of future events. If the scenarios are well grounded in plausible futures, the scenarios should provide a meaningful measure of the demand uncertainty.

Knowing how uncertain the future is benefits the planner in the sense that the greater the uncertainty, the more the planning process is required to develop strategies that offer flexibility and protection against that uncertainty. Such strategies might include some or all of the following:

- demand management;
- fuel diversification;

- smaller unit sizes for central station generation;
- small dispersed generation;
- development of alternative generation technologies.

To conclude, we would like to provide an analysis of uncertainty using scenarios and a more complete answer to the questions posed earlier concerning the rate of electricity demand growth. We have developed four alternative economic and electricity demand growth scenarios for New York State, which are summarized in the following paragraphs. It is hoped that they will provide food for thought.

The rebound (i.e., high-growth) scenario depicts a resumption of the pre-1970 overall growth patterns for both the national and New York State economies, although the specific driving forces differ from those in the past. GNP advances at an average rate of 3.8 percent, close to the economy's potential, and GSP advances at 2.9 percent. This would result in a growth rate for electricity on the order of 2.7 percent, more than twice the current forecast. Thus, capacity constraints would occur by the early 1990s.

The low-lag scenario assumes a lower-than-historical U.S. growth of 2.0 percent annually. Inflation is volatile. Energy prices are high and act as a drag on economic growth. The scenario is essentially an instant replay of the decade 1973–83. In this context of a slower-growth national economy, growth in New York State averages only 0.9 percent a year, which would result in a growth rate in electricity demand of 0.8 percent a year. Thus, existing capacity, including units under construction, would be sufficient into the next century. This scenario produces a forecast for electricity demand that is of the same order of magnitude as the current NYPP planning forecast. The electricity demand forecast for those two scenarios as opposed to the current NYPP and SEO forecasts is summarized in Fig. 6.4. These two scenarios could be characterized as the so-called high/low scenarios.

Finally, Fig. 6.5 presents the electricity demand forecast versus forecasts of the NYPP and the SEO that would evolve from the other two scenarios. These two scenarios, which could be characterized as the middle-ground scenarios, can be described as:

1. A continuation scenario, which develops a forecast based on the growth rate actually experienced between 1974 and 1984 (i.e., the post-oil-embargo era);
2. A NYMOD scenario, which develops a forecast based on the September 1984 economic forecast for New York from the New York Economic Model, published by Wharton Econometric Forecasting Associates.

The differences between the four scenarios, the SEO forecast, and the current NYPP planning forecast for New York State are quite profound. According to a conservative middle-ground view of the future, the capacity of the present system, including units under construction, is inadequate. Shortages may be expected in the 1995–96 period with subsequent adverse effects on New York's economy and electric power supply system.

FIGURE 6.4   Rebound and Low-Lag Scenarios versus SEMP and NYPP Projections

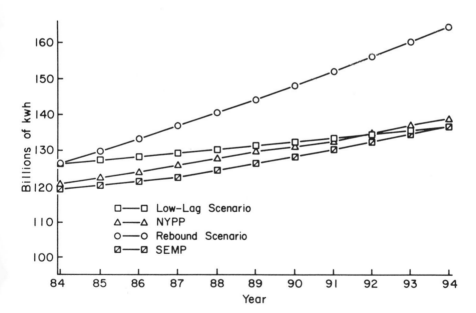

FIGURE 6.5   NYMOD and Continuation Scenarios versus SEMP and NYPP Projections

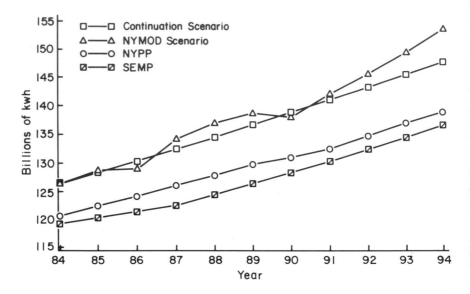

7

Determinants of the Demand for Electricity in New York: Economic Conditions, Nuclear Power Costs, and Primary Fuel Prices ——————

Timothy D. Mount
William J. Deehan

Expectations about the future need for electric power in the United States have changed dramatically over the past ten years. Prior to the oil embargo in 1973, the conventional view in the utility industry was that the demand for electricity would increase from about 1.6 trillion kWh in 1970 to 10 trillion kWh by the end of the century. Most of this increase would be provided by nuclear power. Demand has grown relatively slowly, however, and 1983 demand was only 2.3 trillion kWh in the U.S.; extrapolating the forecast made in 1984 by the National Electric Reliability Council (NERC) gives 3.7 trillion kWh for the year 2000.[1]

A second issue, which is partly responsible for lowering the expectations of future growth in the utility industry, is that nuclear power has turned out to be an expensive source of electricity. Power from one of the two nuclear plants under construction in New York State (Shoreham) is estimated to cost three times as much as power from the new coal plant (Somerset).[2] This latter cost is itself higher than the average cost of producing power in New York State from existing plants. The problem of high capital costs for nuclear power plants is faced by many utilities, and not just those in the New York Power Pool (NYPP). The full consequences of these high costs on the future price of electricity and on economic growth in the nation have not yet been determined.

In reviewing the history of demand forecasting in New York, we find that the failure of the industry to understand the market for power dur-

This research was supported by the New York State Energy Research and Development Authority and the College of Agriculture and Life Sciences, Cornell University. The authors wish to thank Martha Czerwinski and Kathleen Krause for their contribution to the analysis using the CCMU model, Debra Turck for typing the manuscript, and Joe Baldwin for preparing the figures.

ing the 1970s greatly contributed to the current problem of excess capacity. In addition, changing economic conditions during the 1980s have made current construction projects much less economically attractive. High real interest rates and cost overruns for two new nuclear power plants make capital commitments a major threat to the future growth of demand for power from the central grid.

We conclude that the regulatory treatment of these costs will have a significant effect on the future level of demand that can be expected by the NYPP. Established mechanisms for setting rate schedules for electricity are inadequate in these circumstances. Some modifications have already been considered and no doubt others will be suggested. The final decision on how the costs are distributed will affect not only the competitiveness of electricity with other fuels but also the ability of the state to sustain the current economic recovery.

THE FORECASTING RECORD

For the twenty-five years following World War II, the electric utilities in the U.S. and in New York experienced rapid growth in demand. The conventional forecast during this period was that demand would double each decade. Two factors that were instrumental in this phenomenon were the growth in economic activity and declining real prices of electricity relative to the prices of primary fuels.

The situation changes completely after the oil embargo of 1973. The price of oil increased substantially and there was an economic recession. The term "stagflation" was used to describe the lack of growth coupled with rapid price inflation. The utility sector in New York, being heavily dependent on oil, passed on rising real costs to customers. The result of these changing conditions was that the demand for electricity grew very slowly compared with the situation during the 1950s and 1960s.

For eight consecutive years after the oil embargo in 1973, forecasts of the future demand for electricity made by the NYPP were reduced. This record is illustrated in Fig. 7.1 for energy requirements (the level of generation needed to meet demand).[3] A similar picture would emerge if one looked at the forecasts for the nation made by the NERC. Throughout the 1970s, the growth rates forecast by the NYPP were unrealistically high despite the efforts of many analysts to convince utility forecasters that projected growth rates should be lower.[4] These forecasts reflect a serious misunderstanding of the economic factors that determine the demand for electricity.

One illustration of the range of forecasts of future demand in the NYPP that existed in the 1970s is given in Fig. 7.2. Most of the forecasts

FIGURE 7.1 Forecast of Energy Requirements Made Annually by the NYPP, 1973–1984

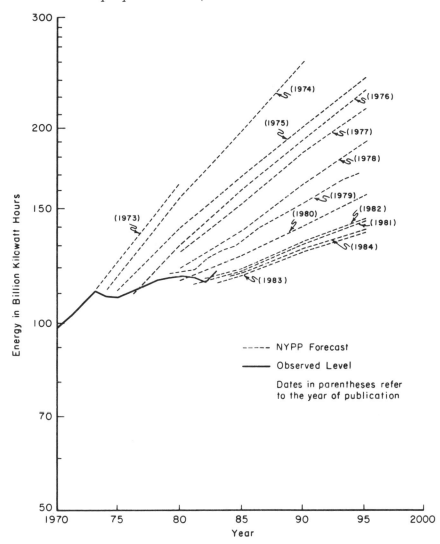

shown were made in 1978, just prior to the publication of the first State Energy Master Plan (SEMP) by the State Energy Office (SEO).[5] In decreasing order of magnitude, the forecasts were made by a) NERA (National Economic Research Associates working for the NYPP); b) NYPP;

FIGURE 7.2   Alternative Forecasts of Energy Requirements for New York State, Published in 1978 and 1979

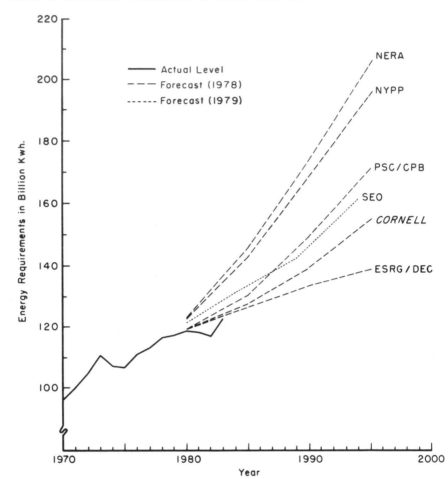

c) PSC/CPB (the Public Service Commission supported by the Consumer Protection Board); d) SEO; e) Cornell University (represented by Caldwell, Greene, Saltzman, and Mount); and f) ESRG/DEC (Energy Systems Research Group supported by the Department of Environmental Conservation).

It is clear from Fig. 7.2 that the two highest forecasts were made for or by the NYPP. The difference between the highest and lowest forecast in 1990 corresponds to approximately 14,000 MW of additional capacity (including a 22 percent reserve). This amount is over seven times the capacity of the two nuclear plants currently under construction (1,889 MW). It should also be noted that the NYPP's recent forecast, shown in Fig. 7.1 (April 1984), would fall between the Cornell and ESRG/DEC forecasts shown in Fig. 7.2.

Actual levels of energy requirements for 1979–83 are also shown in Fig. 7.2. Note that actual levels have been lower than all of the forecasts. The decline in demand in 1981 and 1982 can be attributed to a large extent to the economic recession, but in 1983 demand increased sharply. Current indications suggest that this increase will continue in 1984. An interesting question, which we will address later in this chapter, is whether these increases in 1983 and 1984 are signs of a return to the high rates of growth experienced prior to the oil embargo in 1973.

Since large power plants take many years to plan and build, the high forecasts made in the 1970s by the NYPP resulted in the construction of new generating capacity that has not been needed to meet demand. This is illustrated in Fig. 7.3.[6] The line measuring peak load plus a 22 percent reserve margin represents the capacity needed to provide reliable power to meet demand. At the present time, actual reserve margins are about double those needed to ensure reliability.

Higher reserve margins imply that capital costs must be spread over fewer kWh sales. This increases costs per kWh when the capital charges of new plants outweigh their fuel savings. This in turn lowers sales further. Unfortunately, if traditional regulatory practices are followed, the problem is likely to get worse before it will get better.

The bulk of the cost of the two nuclear power plants under construction has not yet entered the rate base. In addition, standard tax and regulatory practices imply that capital costs charged to customers are much higher in the early years of operation of a new plant.[7] The combined effect of this convention for setting rates and high construction costs has led to use of the term "rate shock" to describe customers' likely response to the use of existing regulatory practices for allocating the costs of the new nuclear plants. The overall result would inevitably be higher prices and lower levels of demand than would otherwise be the case if construction plans had remained in balance with actual levels of demand. It should be no surprise that the final decision on how and to what extent costs of constructing the new plants are passed on to customers will have a substantial effect on the future demand for electricity. The effect will be both direct, by influencing the competitive position of electricity relative to primary fuels in the state, and indirect, by influencing the competitive ability of the state to attract new industries and keep existing ones.

FIGURE 7.3 Capacity Requirements for New York State

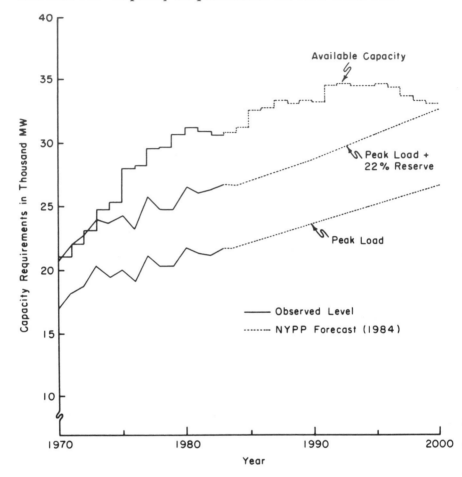

THE CCMU MODEL

The model used in 1978 to derive the Cornell forecast of energy requirements in the NYPP was a relatively large econometric model (600 equations) of New York State's economic and demographic characteristics.[8] The electric utility sector formed only a small part of this model, and it contained little detail about the regulatory practices used to determine prices for electricity. One advantage of the final version of the model was that the economic, demographic, and energy components were fully integrated. Consequently, it was possible to assess how higher energy prices affected not only the demand for energy but also levels of employment, income, and the size and composition of the population.[9]

More recently, an advanced utility simulation model (AUSM) has been developed by the University Research Group on Energy (URGE).[10] The group includes researchers from Cornell University (CU), the University of Illinois at Champaign-Urbana (UCU), and Carnegie-Mellon University (CMU). As the name implies, the model simulates the operation of a utility system, and it operates on an annual basis using a state as the regional unit. The development of the URGE/AUSM was funded by the U.S. Environmental Protection Agency as part of the research program on acid rain. Consequently, an important feature of the model is that it is designed to assess the effects of different policies on the emission of sulfur and nitrous oxides from individual power plants.

During the past year, the New York State Research and Development Authority has funded a project at Cornell to develop an AUSM for the NYPP. This is the model that will be used in the section that follows to derive forecasts. The model is called the CCMU model because it contains components developed at Cornell and Carnegie-Mellon University. These components are very similar to those used in the URGE/AUSM, but the outputs from some components, such as capacity planning, are treated as exogenous inputs rather than being determined within the model.

The structure of the CCMU model is summarized in Fig. 7.4. The four major components are a) demand, b) dispatch, c) pollution control, and d) finance. Information about the cost and quality of fuels, and the construction plans for new generating facilities are provided exogenously.[11] The four components are integrated so that if tighter controls on emissions occur, the higher cost of producing electricity will be reflected in higher rates charged to customers. Higher rates will then reduce demand and determine the appropriate revenue received by utilities. This annual feedback mechanism represents the important interaction of supply and demand forces in the market. It should be noted that many utility models do not operate in this integrated manner, and as a result it is then difficult to assess the effects of different policies on the revenues received. While demand is relatively unresponsive (inelastic) to price in the short run, this is not the case in the long run, and it is the long run that matters for planning purposes.

At the present time, modifications to the CCMU model are still being made. In particular, a revised version of *dispatch* is being developed that will improve the representation of the load duration curve, the pricing mechanisms for imported hydro power, and the choice of fuel by sulfur content at specified plants. Consequently, the forecasts derived in the

FIGURE 7.4 CCMU Model

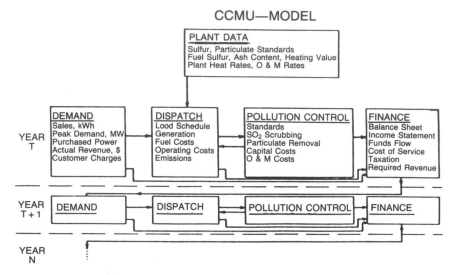

CCMU—MODEL

PLANT DATA
Sulfur, Particulate Standards
Fuel Sulfur, Ash Content, Heating Value
Plant Heat Rates, O & M Rates

YEAR T

DEMAND
Sales, kWh
Peak Demand, MW
Purchased Power
Actual Revenue, $
Customer Charges

DISPATCH
Lood Schedule
Generation
Fuel Costs
Operating Costs
Emissions

POLLUTION CONTROL
Standards
SO₂ Scrubbing
Particulate Removal
Capital Costs
O & M Costs

FINANCE
Balance Sheet
Income Statement
Funds Flow
Cost of Service
Taxation
Required Revenue

YEAR T+1

DEMAND — DISPATCH — POLLUTION CONTROL — FINANCE

YEAR N

Source: Adapted from Figure 3 in Chapman, *et al.,* "Air Pollution, Nuclear Power and Electricity Demand: An Economic Perspective," A.E. Res. 83-34, Department of Agricultural Economics, Cornell University, September 1983.

following section using the CCMU model should be treated as preliminary. However, the qualitative aspects of the results are not expected to change.

One final qualification: the CCMU model represents the utility sector only, and it is not integrated into a comprehensive model of New York's economic and demographic characteristics.[12] Economic and demographic variables are used as inputs into the CCMU model, and as a result, there is no feedback mechanism for assessing how higher electricity prices influence economic growth in the state. Given the importance of the cost implications of nuclear power, this lack is unfortunate. Reestablishing a fully integrated model with economic, demographic, and energy components would be a sensible objective for future research.

FUTURE PROSPECTS OF THE NYPP

In order to use the CCMU model for forecasting purposes, assumptions about economic growth and fuel prices must be specified for the forecast period. Additional information about future changes in installed capacity must also be supplied. Three different scenarios are defined; in addition, the first two are used to evaluate a number of specific changes in the input assumptions to illustrate the likely effects of construction plans currently proposed in New York. These plans include converting existing oil-burning plants to coal, as well as completing new generating facilities.

The characteristics of the three scenarios are summarized in Table 7.1, and a more detailed description is given in the appendix. The base and alternative scenarios provide forecasts for the period 1985–99, and WISHB (the "way it should have been") for the period 1980–99. For the base scenario, fuel price assumptions are developed from recent discussions with staff members at the SEO and imply that real oil prices will continue to fall during the 1980s and then increase throughout the 1990s. In addition, the current economic recovery is sustained throughout the forecast period. Three new generating plants are completed (Somerset, Shoreham, and Nine Mile Point Two) and four oil plants are converted to coal (Arthur Kill, Ravenswood, Danskammer, and Lovett).

The forecast of energy requirements for the base scenario is shown in Fig. 7.5, together with the forecast made earlier this year by the NYPP. The NYPP's forecast is lower than the base. This is largely a result of the predicted drop in demand in 1984 forecasted by the NYPP; otherwise the implied growth rates after 1985 are similar for the two forecasts. As stated earlier, it appears at this time that demand in 1984 will be at least 3 percent higher than it was last year.

One important feature of the base forecast is the declining growth rates exhibited from 1985 to 1990. These reflect the influence of passing the full cost of coal conversions and of constructing the two new nuclear plants, particularly the latter, into the rate base. Even though the CCMU

TABLE 7.1 Summary of the Input Assumptions Used to Define the Scenarios

	Base	WISHB	Alternative
Forecast period	1985–99	1981–99	1985–99
Economic	moderate growth high real interest	moderate growth low real interest	slow growth same as base
Real oil prices			
1985–89	slow decline	moderate increase	same as base
1990–99	moderate increase	moderate increase	same as base
Real coal prices			
1985–89	fast increase	moderate increase	same as base
1990–99	level	moderate increase	same as base

FIGURE 7.5 Forecasts of Energy Requirements for New York State

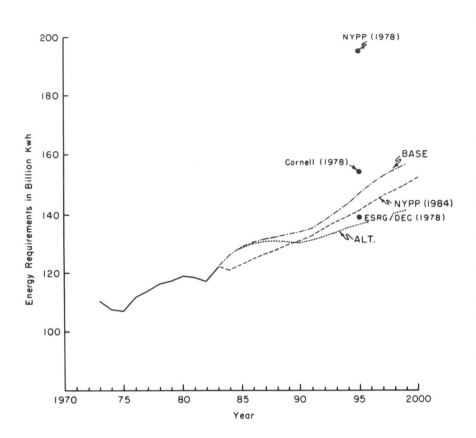

model operates at the state level, the costs are large enough to have a sub-stantial effect on aggregate demand.[13] This condition reflects the use of standard regulatory procedures for passing on capital costs, and conse-quently, large rate increases are associated with the addition of any ex-pensive new power plant into the rate base. Growth rates increase gradu-ally after 1990 because the capital costs added to the rate base are depreciated over time, construction-related tax benefits from earlier years are passed back to customers, and the cost of fossil fuels is reduced by operating the nuclear plants and substituting coal for oil. In other words, if the new plants continue to operate for long enough, the average cost of generation will eventually be lower than it would otherwise have been without the plants.[14]

The adverse effects of the new plants on demand in the early part of the base forecast have been exacerbated by a number of factors that were not anticipated by the utilities. In particular, high real interest rates have increased costs through the allowance for funds used during construction. Delays are expensive when real interest rates are so high. In addition, fuel savings from using nuclear power or coal are smaller because oil prices did not increase as fast as expected after 1981 and are, in fact, currently falling in both nominal and real terms. For both of these reasons, the time needed for a new plant to pay for itself is longer than it would have been had conditions not changed so dramatically.

The second scenario, WISHB, is designed to represent expectations that existed for economic growth and fuel prices during the hearings for the 1979 SEMP. Lower real interest rates and continuous growth of both employment and fuel prices from 1980 to 1999 characterize this scenario. The base and WISHB scenarios are used to compare the effects of current construction plans on the cost of power (using traditional rate-making procedures) under two different sets of economic conditions. It is interesting to note that the 1979 vintage assumptions in WISHB result in a level of demand that is only slightly higher than the base scenario in 1999 (158 versus 156 billion kWh). This outcome demonstrates that the positive effect of economic growth on demand can be offset by higher electricity prices and that the use of the WISHB assumptions would not result in the high growth forecast at that time by the NYPP.

In 1979, the major objectives of the long-range construction plan of the NYPP were to add capacity to meet anticipated increases in demand (see Table 7.2), and to substitute for oil capacity to reduce New York's dependence on imported oil. Consequently, proposals were adopted for constructing new plants, completing the plants under construction, and for converting a number of existing oil plants to coal burning. Since demand has not grown as fast as expected, more emphasis has been put on saving oil as the justification for continuing with current construction plans.

In an effort to evaluate these decisions a posteriori, the model was also run for the base and WISHB scenarios again, but this time a) without going ahead with coal conversions and b) without constructing the two new nuclear facilities. In Table 7.2, the average cost of service in real terms is presented in 1990 and 1995 for the three cases (all construction completed; no coal conversions; no new nuclear plants) using the two different scenarios (base; WISHB).

For the base scenario, costs are lower in 1990 and higher in 1995 if no coal conversions are made and if no nuclear plants are built. This out-

TABLE 7.2 Average Cost of Service Using Different Construction Plans (Preliminary Results)

Case	Average cost (1980¢/kWh)				Difference from case 1 (%)				Switch year*	
	Base		WISHB		Base		WISHB		Base	WISHB
	1990	1995	1990	1995	1990	1995	1990	1995		
1. All construction completed	6.50	5.64	5.82	6.20						
2. No coal conversion	6.40	5.75	5.94	6.47	−1.5	+2.0	+2.1	+4.4	1993	1988
3. No new nuclear	6.23	5.71	5.84	6.36	−4.2	+1.2	+0.3	+2.6	1993	1990

*First year in which the cost of service is higher than the corresponding level in Case 1.

come demonstrates that both types of investment eventually reduce the cost of power. The initial effect of adding these capital costs to the rate base using standard regulatory procedures is to increase the cost substantially for a number of years. The relatively small difference in costs between cases 1 and 3 in 1990 reflects the fact that the nuclear plants have already been included in the rate base for a number of years and that the cost is an average for the whole state, not just for the utilities building the new plants.

In addition to the cost of service, Table 7.2 shows the first year in which the average cost is lower than it would be without the investment in coal conversions or in new nuclear plants. Under the base scenario, the first year when costs are lower is 1993 in both cases. In other words, 1993 is the first year in which benefits (lower costs) start paying off the higher costs incurred in earlier years.[15] With real discount rates of 7 percent in the base scenario, the present value of a dollar of benefits in 1993 is only $0.54 today. In today's economic environment, it will take many years for the present value of future benefits to equal the present value of near-term construction costs, and it is possible that this can never be achieved.[16]

Comparing the results for the base and WISHB scenarios, two related differences are apparent. First, the long-run cost savings from coal conversions and new nuclear plants are greater with WISHB. This is demonstrated by the larger percentage reductions in cost shown for 1995 with WISHB. Second, the first years in which net benefits are positive occur earlier with the WISHB scenario than with the base scenario for both types of investment. This reflects the lower capital costs associated with lower real interest rates in WISHB, and the greater savings from reducing the use of oil for generation because oil prices do not fall in this scenario.

The overall purpose for comparing the WISHB and base scenarios is to illustrate the extent to which changing economic conditions between 1980 and 1984 have adversely affected the economic viability of current construction plans. Unlike the four coal conversions, a large proportion of the total cost of the two new nuclear plants has already been committed. Under current economic conditions, the urgency for converting oil plants to coal is no longer present.

GENERATION COSTS AND ECONOMIC GROWTH

The high costs associated with the addition of two nuclear plants into the rate base using standard regulatory procedures have a substantial dampening effect on the growth of demand in the base forecast (see Fig. 7.5). Despite the size of this effect, the forecast probably underestimates the actual response that is likely to occur. First, the structure of the CCMU model assumes that the higher costs are borne on a statewide basis. Based on current regulatory practices, however, cost increases will be much higher for some utilities. Second, there is no feedback mechanism between energy prices and economic activity as there is in our model of the New York economy. Consequently, income and employment levels are not affected by higher electricity prices in the current model. It is, however, possible to draw some inferences about these interrelationships from earlier analyses.

A number of studies conducted at Cornell have investigated the relationships between energy prices and economic activity. Two in particular are closely related to the topic of this chapter. The first study investigated how different policies for pricing fuels and electricity affect employment and income.[17] The model employed for this analysis was developed from the one used to derive the Cornell forecast in Fig. 7.2. The second study focused on the pricing and allocation of hydro-power, and considered the implications of alternative pricing schemes on the primary metals and chemical industries as well as on total employment and income in the state.[18] Both studies investigated the consequences of higher electricity prices on the state's economy.

In the two analyses cited above, production levels in manufacturing are inversely related to the cost of production in the state relative to the costs in the rest of the nation. In other words, if costs increase in the state relative to the nation, an estimate is made of how much production and employment levels in manufacturing will fall. The general conclusion from the two studies is that charging higher prices for energy will reduce total employment levels and income in the short run. In the long run (five years), it is possible that additional production and employment in the commercial sector may compensate for losses in manufacturing, so that income levels can actually be higher with higher energy prices. This only happens, however, if there is some source of compensation for higher energy prices in the state. Taxing oil consumption and using the revenues to reduce income taxes is one example of a policy that increases the price of energy but also stimulates the economy by giving people higher disposable incomes. In contrast, deregulating the price of natural gas has no compensating effect for higher prices because the profits are transferred out of the state. As a result, the former policy leads to higher incomes in the long run, while the latter implies lower incomes.

The general conclusion from these studies is that higher energy prices in the state do have an adverse effect on economic activity, particularly in the manufacturing sector. Growth in other sectors can compensate for this adverse effect only if higher prices are coupled with higher disposable income in the state. Compensating effects will not occur with higher prices for electricity associated with the addition of new nuclear plants into the rate base. In such a situation, a large part of the higher expenditure on electricity would be transferred to stockholders and creditors of the utilities. Only a small fraction of this transfer is likely to be spent in the state and, in particular, within the service territories of utilities affected by higher costs. As a result, compensating effects for higher prices will be small, and income and employment levels will almost certainly be reduced in service territories subjected to substantially higher rates for electricity. Furthermore, if a company decides to leave a particular region of the state because of high electricity prices, it is quite likely that a decision would be made to locate in another state or country rather than in another region of New York.

The third scenario described in Table 7.1 (the alternative scenario) is designed to reflect a less optimistic set of assumptions about economic growth in the state based on what would happen if much higher rates were charged, to pay for the full cost of new nuclear plants. Although positive economic growth is assumed, employment in manufacturing continues to decline slightly in this scenario, as it has since 1980. All other exogenous assumptions are the same as in the base scenario.

The forecast for the alternative scenario is shown in Fig. 7.5. The declining growth rates of demand from 1985 to 1990 are still evident. In fact, demand actually decreases in 1988 and 1989 even though current growth rates are relatively high. The final level of demand forecast for 1999 (142.5 billion kWh) is well below the base forecast (156.5 billion kWh) and the current forecast of the NYPP (149.8 billion kWh). It is important to note that the difference between the forecasts in the base and alternative scenarios is attributable to different economic assumptions only. We do not intend that the alternative scenario be interpreted as a projection of the most likely level of future demand; rather, it describes what would happen if electricity prices increase substantially relative to the prices of primary fuels. In these circumstances, economic growth is no guarantee that the demand for electricity will grow as fast as it would with lower prices.

The conclusion is that the indirect effects of higher electricity prices on economic growth can reduce the growth of demand for electricity. Since price increases could be very large in some service territories, these indirect effects should not be ignored. However, the size of the price increases will be determined largely by decisions regarding how and to what extent the costs of the nuclear power plants under construction are passed on to customers. Since the exact outcome of these decisions is not yet known, considerable uncertainty about the future demand for electricity exists. It is clear, however, that modifying regulatory procedures to reduce the adverse effects of higher rates is warranted.

CONCLUSION

During the 1950s and 1960s, when the economy was expanding and the real price of electricity falling, the demand for electricity grew approximately twice as fast as the growth in gross state product. The 1970s were a period of slow economic growth coupled with increasing real prices for electricity and primary sources of energy. There was very little growth in the demand for electricity during this period. After a decline in demand in 1981 and 1982, recent growth rates have been higher than at any time since the oil embargo in 1973.

Current expectations for the remainder of the 1980s are that economic growth will continue. If conditions were the same now as they were in the 1950s and 1960s, high economic growth would be associated with high growth in electricity demand. The important difference is that electricity became cheaper relative to primary fuels during the period of high growth in demand. For the next few years, however, the reverse situation is likely to occur. Current projections are that the real price of oil will fall. In contrast, the real price of electricity is likely to increase because of the costs associated with building new nuclear power plants. Under these circumstances, it is very likely that the demand for electricity will grow more slowly than the economy as a whole. For example, with relatively low prices for primary fuels and high prices for electricity, cogeneration will become a more competitive alternative to buying power from the central grid for many customers.

If the full costs of the new nuclear plants are passed on to customers using standard regulatory procedures, large price increases for electricity will occur in the mid to late 1980s. It is shown that these increases will slow the growth of demand substantially, even on a statewide basis. Furthermore, it is argued that such price increases would jeopardize the economic recovery that is currently under way in the state. The implication is that some modifications to existing regulatory practices will be required to avoid these adverse effects. For example, the repayment of capital costs could be spread more evenly over time using economic depreciation as the basis for passing on costs.

A more controversial strategy would reduce the proportion of capital costs that is passed on to the ratepayers. This would lower rate increases and, consequently, would not dampen the growth of demand as much as if the full cost is reflected in rates. For example, if only 50 percent of the cost of Shoreham and Nine Mile Point Two is incorporated into the rate base (using standard regulatory procedures), forecast demand reaches 128 billion kWh in 1990 under the same economic conditions as the base scenario. This represents an increase of 8.0 percent from the forecast level of demand in 1984, compared to an increase of 6.1 percent in the base scenario and only 1.2 percent in the alternative scenario. (The latter scenario reflects the assumption that substantially higher rates for electricity will reduce the rate of growth of the economy as well as the demand for electricity.) In the longer run, the adverse effects of higher capital costs in the short run are offset by lower production costs. However, it is shown in the analysis that the eventual savings in the cost of service are smaller than anticipated, because capital costs are higher than expected and oil prices are lower than expected.

Even if one takes assumptions about economic growth and future fuel prices as given, there is considerable uncertainty about how fast the demand for electricity will grow. Nuclear power is not the only source of uncertainty faced by the NYPP. The cost of implementing emission reductions to meet the objectives of the Acid Deposition Control Act, and the amount of power purchased from Canada will also affect rates and, therefore, future levels of demand in the state. Ironically, the assumptions that are made about the extent and timing of the recovery of nuclear power costs, and whether or not the nuclear plants operate on schedule, play a significant role in determining both the best way to control emissions and how much power should be imported. In other words, appropriate policy actions can only be established when the uncertainties over nuclear power costs are resolved.

NOTES

1. NERC, "Electric Power Supply and Demand 1984–1993," 1984 (refers to the 48 contiguous states).

2. New York *Times*, November 5, 1984. (Note that Somerset is fitted with scrubbers so that emissions of sulfur oxides meet new source performance standards.)

3. Forecasts of the peak load demanded in a year are directly related to the amount of generating capacity needed. Forecasts of energy requirements and of peak load are, however, closely related to each other. Since energy requirements are not as sensitive to climatic variation from year to year as peak load, observed levels of energy requirements provide a better measure for assessing how demand has changed.

4. Duane Chapman and Tim Tyrrell from Cornell were important participants in this effort.

5. These forecasts were incorporated into the public record for the first State Energy Master Plan. Revised forecasts were also made during the hearings by most parties.

6. The levels of available capacity do not account for all imported hydro-power from Canada. Approximately 2,000 MW could be added to available capacity after 1985 to reflect projected levels of imports fully.

7. For a new plant, construction costs are accumulated over time, together with an interest allowance, until the plant is operating. At that time, the full accumulated cost is transferred to the rate base, and customers start paying the allowed rate of return on this cost, which is then depreciated over a specified number of years to zero.

8. S. Caldwell et al., "Forecasting regional energy demand with linked macro/micro models,"*Papers of the Regional Science Association XXV Annual Meeting* 43 (1980):100–13.

9. T. Considine and T. Mount, "A regional econometric analysis of energy prices and economic activity," *Environment and Planning A* 15 (1983):102–41.

10. J. Stukel et al., *The State-Level Advanced Utility Simulation Model: Program Documentation*, University of Illinois, September 1984.

11. The major components of the model and the persons responsible for the development are: dispatch (S. Talukdar and N. Tyle from CMU); control technologies (E. Rubin, J. Molberg, C. Bloyd, J. Skea, CMU); New York plant and fuel data (G. Fry, CU); demand (T. Mount, M. Czerwinski, CU); finance model (D. Chapman, K. Cole, M. Slott, M. Younger, CU).

12. The demand for primary fuels is estimated in the model, however, so that the effects of interfuel substitution can be determined.

13. This conclusion is consistent with earlier results using the CCMU model presented in L. D. Chapman et al., "Air Pollution, Nuclear Power and Electricity Demand: An Economic Perspective," A.E. Res. 83-84, Department of Agricultural Economics, Cornell University, September 1983. The direct long-run price elasticities for electricity are -.30, -.65, and -.55 for the residential, commercial, and industrial sectors, respectively (see Chapman et al., ibid., Table 8, p. 30). It should be noted that the form of the demand equations used in the model implies that the price elasticities are not constant (in accordance with economic theory), and the values shown are representative levels for the state (see "The State-Level Advanced Utility Simulation Model: Analytical Documentation," chap. 2, "The Demand Model," sec. 2.2.2.1, by T. D. Mount, URGE Project Office, Public Policy Program, College of Engineering, Univ. of Illinois at Urbana-Champaign, 105 Observatory Building, 901 South Mathews Ave., Urbana, IL, 61801).

14. This statement does not consider the costs of decommissioning the nuclear plants or of storing the used nuclear fuel.

15. No attempt is made here to subdivide the cases further. It is highly likely, however, that the costs and benefits vary substantially among plants. For example, converting Danskammer to coal is probably the most cost-effective of the four conversions considered.

16. A thorough analysis of the costs and benefits of these scenarios would discount the real resource costs of construction and of operation of the whole power pool and the social surplus generated in each scenario.

17. T. J. Considine and T. D. Mount, "Energy Pricing, Employment and Economic Growth: An Econometric Analysis of the New York State Economy," A.E. Res. 84-13, Department of Agricultural Economics, Cornell University, July 1984.

18. T. W. Hertel and T. D. Mount, "The Economic Effects of Reallocating Publicly Owned Hydropower in New York State," A.E. Res. 84-7, Department of Agricultural Economics, Cornell University, April 1984.

APPENDIX: SCENARIO DESCRIPTIONS

Base Assumptions

Economic-demographic (Annual growth rates 1984–99, in percent)

Population	0.5	Personal income	2.0
Employment,		Inflation	6.0
manufacturing	0.0	Level of interest rate,	
Total employment	2.0	average	12.0

Electricity supply

Canadian imports, 1984–87: 21 GWh; 1987–97: 24 GWh; 1997–99: 21 GWh

Fuel availability: Natural gas unavailable for boiler fuel after 1986

Increased pollution control
measures — None (current special limitations in effect)

Coal conversions — Ravenswood 3, 1991; Arthur Kill, 1988; Danskammer, 1989; Lovett, 1986.

New production plants
included — Somerset, Nine Mile Point Two, Shoreham

excluded — Prattsville

Electric-generating fuel price (Real annual growth, in percent)

	1984–87	1987–90	1990–97
Oil	−0.45 to −0.85	−0.45 to −0.85	2.08 to 2.21
Coal	4.60 to 6.50	0.00 to 2.60	0.00
Natural gas	−0.45	N/A	N/A
Nuclear	0.00	−1.54	−2.37

WISHB Assumptions

Economic-demographic (Annual growth rates 1984–99, in percent)

Population	0.5	Personal income	2.5
Employment, manufacturing	3.0	Inflation	10.0
Total employment	1.5	Level of interest rate	12.0

Electricity supply

Canadian imports	(Same as Base)
Fuel availability	''
Incremental pollution control measures	''
Coal conversions	''

New production plants
    included   (Same as Base)
    excluded   ʺ

Electric-generating fuel price (Real annual growth, in percent)

	1984–87	1987–90	1990–97
Oil	3.2	2.8 to 3.2	2.8
Coal	2.0	2.0	2.0
Natural gas	6.5	N/A	N/A
Nuclear	0.0	0.0	0.0

Alternative Assumptions

Economic-demographic (Annual growth rates 1984–99, in percent)

Population	0.5	Personal income	1.0
Employment, manufacturing	−0.5	Inflation	6.0
Total employment	2.0	Level of interest rate	12.0

Electricity supply

Canadian imports   (Same as Base)
Fuel availability   ʺ
Incremental pollution control
   measures   ʺ
Coal conversions   ʺ
New production plants
   included   ʺ
   excluded   ʺ

Electric-generating fuel priced (Real annual growth, in percent)

	1984–87	1987–90	1990–97
Oil	(Same as Base)	(Same as Base)	(Same as Base)
Coal	ʺ	ʺ	ʺ
Natural gas	ʺ	ʺ	ʺ
Nuclear	ʺ	ʺ	ʺ

8

The World Oil Market and New York Electricity

Douglas R. Bohi
Joel Darmstadter

Within our broad subject, electric utilities and their statewide and nationwide sales to the industrial sector of the economy, we have selected for particular attention two key sets of questions. *First*, how have developments on the world oil scene, combined with other factors, shaped the emerging industrial electric power market? Nothing resembling the international energy upheavals of the 1970s has ever been experienced before. A dissection of the process by which these events affected domestic fuel and power costs might provide an instructive backdrop to thoughts about the prospects for reverberations from recurrent oil shocks. Significant though the shock waves transmitted via international oil markets may have been, it is also worthwhile to look at additional changes affecting the electric power markets in recent years. *Second*, given the unfolding state of the world oil market, the lessons of the past decade, and economic adjustments taking place both in industry and utilities, what is the prospective resilience to any future oil price gyrations or supply disruptions? Even while avoiding making a forecast, we believe useful things can be said about the factors likely to influence and constrain future oil prices.

The institutions on which we will largely concentrate are the industrial enterprises that use electricity and other energy forms in their production activity, and the utilities that generate the electric energy distributed to those firms.

ENERGY, ELECTRICITY, AND NEW YORK* INDUSTRIES

We can begin with a few basic facts, although there is probably little reason to belabor those familiar features about the New York energy

---

*Unless otherwise specified, "New York" refers to the state of New York.

situation that predispose the state to external energy shocks. New York utilities, proportionately much more than those in the country as whole, rely on petroleum as their major fuel source (see Table 8.1). Thus, even for industrial firms sheltered from the *direct* impact of oil price shocks, such effects are transmitted indirectly through increases in industrial *electricity* prices. Nonetheless, for reasons which will be discussed below, the share of electricity in both statewide and U.S. industrial energy consumption has grown markedly (see Table 8.2).

Another fact of life relates to New York's fuel and power costs—traditionally these exceed average national levels by a considerable margin (see Table 8.3). However, contrary to the differences in price *levels*, OPEC-driven energy price *changes* during the last decade have been no kinder to the rest of the country than to New York: both experienced unprecedentedly large increases, though the rate increases posted by some severely affected oil-burning utilities, such as ConEd and the Long Island Lighting Company (LILCO), were particularly sharp. These increases, which carried other fuel costs along in their wake, were only mildly attenuated by the fact that the overall price index was itself moving up at a rapid rate. In other words, *real* prices rose steeply for all energy forms, though more for fuels than for electricity.

Oil Price Links between New York and the World

The process by which energy price rises—spearheaded by OPEC oil—become generalized is tied in with the fact that energy forms are sub-

TABLE 8.1 Electricity Generation by Energy Source (Percent)

Energy Source	United States		New York State	
	1972	1982	1972	1982
Coal	42.2	51.7	12.8	13.1
Gas	22.0	14.0	6.6	12.5
Petroleum	16.7	6.5	48.5	28.0
Hydro	15.7	14.7	26.1	33.6
Nuclear	3.2	12.9	6.0	12.6
TOTAL[a]				
	100.0	100.0	100.0	100.0

Source: United States Department of Energy/Energy Information Administration (DOE/EIA), *State Energy Report: Consumption Estimates, 1960–1982,* May 1984.
[a]Includes small amounts of other sources not shown separately.

TABLE 8.2 Electricity's Share of Industrial Energy Consumption (Percent)

	United States	New York State
1960	6.6	6.6
1970	8.4	9.1
1972	9.0	10.5
1977	11.1	11.9
1980	11.8	15.5
1982	12.9	19.5

Source: DOE/EIA, *State Energy Report: Consumption Estimates, 1960–1982*, May 1984.

TABLE 8.3 Fuel and Power Costs ($/Million BTUs)

	United States		Annual Percent Change	New York State		Annual Percent Change
	1972	1981		1972	1981	
Industry	0.88	5.61	23	1.23	7.00	21
Fuels	0.64	4.73	25	0.85	5.05	22
Electricity	3.41	12.48	16	4.18	15.38	16
Utilities[a]	0.42	2.00	19	0.59	3.25	21
Coal	0.38	1.53	17	0.51	1.71	14
Gas	0.33	2.79	27	0.46	3.49	25
Oil	0.66	5.42	26	0.66	5.12	26

Source: DOE/EIA, *State Energy Price and Expenditure Report 1970–1981*, June 1984.
[a]Including purchases of nuclear and other fuels not shown separately.

stitutable among each other in a wide range of industrial processes and end-use activities, and through the mix of products that people consume. Such substitutability is possible for a given installation or piece of industrial equipment; or it can be exploited, over the long run, as one fuel-using asset is replaced by another.

With respect to short-run substitutability among energy forms, half of New York's oil-based utility generating capacity can be switched to gas. When it comes to New York industrial establishments, we lack information about energy-input flexibility. (Nationally, about 70 percent of fuel oil used in large manufacturing plants is estimated to be replaceable through alternate combustion capability.) We do have some notion about such flexibility for the Middle Atlantic region as a whole (which includes New Jersey and Pennsylvania, along with New York). The picture for the region reveals something of an anomaly. For, despite being the part of the country with historically the largest degree of oil-import dependence, the region shows considerably less industrial fuel-switching capability than a number of other areas. Part of the explanation may arise from the technical features of endemic manufacturing processes, which have precluded significant fuel switching. But another part of the explanation may lie in the term "historically." At a time before the security of the oil supply became a perceived problem, the Northeast enjoyed the benefits of cheap residual fuel oil imports that were free of quotas. And even though the 1973–74 jolt undermined prior perceptions, the area's oil dependence was cushioned by the entitlements system; concurrently, the attractiveness of laying in gas-firing capacity as an alternative to oil was undermined by a growing unreliability of gas supplies—then beginning to be increasingly diverted to intrastate markets in places like Texas and Louisiana. When one takes account of these factors, and the inertia governing decisions about the acquisition of capital stock, lags in fuel-switching investments seem less puzzling. Still it would be helpful to know whether past and prospective oil-price shocks have intensified industrial fuel-flexibility measures in this part of the country.

Fuel substitutability characteristics help to determine the extent to which oil-market perturbations transmit their effect onto the broader configuration of fuel and power prices faced by U.S. energy consumers. Fuel use flexibility can mute the severity of an oil price increase, although, conversely, the shift in demand to alternate energy forms will drive up their prices. The magnitude of the generalized energy price response will also depend on factors other than fuel switching—for example, environmental regulations or the consequences of increased demand for U.S. coal exports. Then, too, the amount of fuel flexibility may turn out to be excessive if it is assumed that supplies of alternative fuels will be available.

The Changing Energy Mix for Utilities and Industry

Having sketched out the process by which other energy prices were affected by oil price increases, we briefly consider the responses by utilities and industry (apart from installing multiple-fuel flexibility, discussed above) to the new fuel and power price structure. It goes without saying that rising energy prices have restrained energy demand growth across a wide swath of the economy in New York and throughout the nation. But, on a more specific level, how have *relative* price movements dictated energy choices by utilities and particular industries? In considering this, it must be kept in mind that electric power has become a better buy than primary fuels for industry; and that, from a narrow input cost standpoint, coal should have become the fuel of choice in electricity generation.

The utility story is probably easiest to dispose of. Fuel choices for New York utilities are governed by factors that are not nearly as important for the country as a whole as they are for the state. Environmental regulations inhibit the use of coal. (We are told that the contemplated large-scale conversion of existing oil units to coal—to the point where coal would account for nearly 70 percent of the incremental kilowatt-hours generated to the year 2000—is likely to conflict with recently enacted acid rain control statutes. Hydro-power from Quebec might mitigate that problem; but substituting imports from coal-burning utilities in Ontario could just shift the scene of the crime but not the nature of the environmental offense.) In addition, New York State capacity additions have been, and will continue to be, modest. The burden of shifting away from oil is imposed on existing plants rather than via the easier route of newly constructed facilities. Thus, it is not surprising that the U.S. as a whole has generally been more successful at oil and gas changeover policies than the state. Eventually, however, Congress will have to address the question of whether the remaining statutory provision of the Fuel Use Act—barring use of oil and natural gas in new utility and large industrial boilers—is consistent with the circumstances of a state like New York.

Changes that have occurred during the previous decade in the industrial mix of electricity relative to the other energy forms are less well explained by regulatory factors than economics and technology. (Conversion to coal, originally mandated by the Fuel Use Act, has been waived on grounds of impracticability.) As we saw earlier in Table 8.3, the price of electricity has declined relative to other fuel prices, and its share of the industrial energy market has grown both in New York and around the country. A more specific look at the manufacturing portion of the (more broadly defined) industrial segment of the economy helps bring this out (see Table 8.4).

Numerous manufacturing industries raised the electricity share of their energy consumption during the 1971–81 period. And the census statistics reveal a few anomalies: traditionally nonelectrical machinery, an important state industry, shifted to electric power in a big way; while printing and publishing, an even more important state industry, reversed direction after recording a big leap in its electricity share in 1977 (not shown in Table 8.4). (A statistical oddity must be noted: the electricity share in the industrywide Department of Energy/Energy Information Agency [DOE/EIA] data set for Table 8.2 gives much smaller percentages than the census figures for manufacturing only in Table 8.4. Even though the two data series are compatible with respect to general trends, a precise reconciliation would be helpful.)

An additional bit of insight is provided by Table 8.5. The tabulation tracks trends in electricity and total energy use per dollar of value added. The electricity/value-added ratio has gone down for a number of

**TABLE 8.4 Electricity as Share (Percent) of
Total Energy Use, Selected Manufacturing Industries,
New York State, 1971 and 1981**

	1971	1981
Primary metals	28	39
Chemicals	17	15
Electrical equipment	33	36
Paper/paper products	12	14
Nonelectrical machinery	20	32
Stone, glass, clay	9	12
Food/food products	13	15
Transportation equipment	26	32
Fabricated metal products	20	24
Rubber/plastics	24	36
Printing/publishing	37	32
Total manufacturing, New York State	20	23
Total manufacturing, United States	14	20

Source: Bureau of the Census, *Census of Manufacturing,* various volumes.

manufacturing industries, while electricity's *share* in total energy, shown in Table 8.4., has risen. However, for nonelectrical machinery, whose operations (as just mentioned) have become more electrified, the industry's electricity use per dollar of value added went down; but the ratio of total energy consumption to value added went down even further. The picture tallies with the notion that the pervasive rise in energy costs reduced overall energy intensity; but the differential magnitude of power and fuel cost increases induced a shift—or, more cautiously, probably contributed to a shift—from primary fuels to generated power.

It turns out that improvements in overall energy efficiency for a given industry tend to be significantly related to the growth of value-added for that industry. Expanded manufacturing activity facilitates the kind of energy-saving investments that stagnating conditions inhibit or make less attractive.

Independent Trends in Industrial Electricity Use

We have been focusing on changes in energy and electricity markets traceable to OPEC-instigated price shocks. But of course other forces may be affecting electricity trends and patterns. In fact, it seems clear that there are some fundamental market and technological phenomena that cannot be overlooked by those concerned with regional energy and eco-

TABLE 8.5 Energy/Value-Added Ratios, Selected Manufacturing Industries, New York State (Thousands of BTUs per 1971 Dollar)

	1971		1981	
	Electricity	Total Energy	Electricity	Total Energy
Primary metals	24.1	86.7	34.1	86.8
Chemicals	5.9	35.0	5.0	32.7
Electrical equipment	3.7	11.0	2.8	7.8
Paper/paper products	8.3	70.2	9.4	66.2
Nonelectrical machinery	2.3	10.9	1.9	6.1
Stone, glass, clay	6.4	67.6	9.1	73.1
Food/food products	2.8	21.9	2.5	16.5
Transportation equipment	2.8	10.8	3.2	10.0
Fabricated metal products	3.6	17.7	2.9	11.9
Rubber/plastics	6.4	27.4	6.2	17.7
Printing/publishing	1.1	3.1	0.7	2.1
Total Manufacturing	3.5	18.0	3.3	14.4

Source: Bureau of the Census, Census of Manufacturing, various volumes; value-added for 1981 is a rough estimate.

nomic growth issues. For example, we can expect to begin seeing interregional convergence in electricity rates as high cost incremental generating capacity becomes a progressively greater proportion of total capacity, thus tending to neutralize the cost-of-electricity factor in U.S. locational decisions. Thus, the historical cost advantage of such areas as the Pacific Northwest, Tennessee Valley, and the Niagara tier of New York State is inevitably eroding: witness the near stagnation of investments in domestic aluminum smelting capacity—indeed, the absence of construction plans for even a single additional production facility. The U.S. *as a nation* may be in the process of relinquishing its hold on electricity-intensive industry.

Another issue to keep in mind is the extent to which relative energy prices remain an important factor affecting the choice of energy forms in industrial activity. If future price changes for oil and gas are expected to be significantly different from the changes for electricity, one would expect electricity to either enlarge or relinquish its industrial market share. Both prospects, however, depend on the characteristics of the capital stock and on existing and prospective technological processes. Shifts away from electricity and toward other energy forms are probably much more constrained in the manufacturing sector than in other parts of the economy.

Even as avowed an electricity skeptic as Amory Lovins implicitly concedes that many of the functions for which electricity is employed in production offer little opportunity for being replaced by substitute energy forms.[1] He labels 94 percent of industrial electricity use as "obligatorily electrical." After all, many electric motors and electrolytic processes simply do not permit recourse to nonelectrical energy forms; and the small amount of process heat produced by electrical means can no doubt be rationalized on technical and economic grounds.

Given the more likely prospect of fuel price increases exceeding those for power, the outlook for a continuing switch from fuel to power in given or modified technologies is much more plausible: for example, precision heat applications, electric steel making and, more generally, environmental factors and security-of-supply considerations may dictate such a fuel-to-power shift. And, as Tables 8.2 and 8.3 disclosed, electricity's share of industrial energy use has been expanding significantly, particularly in New York, during a period in which increases in the industrial cost of electricity lagged behind the price rise for fuels.

CHANGES IN WORLD OIL MARKET CONDITIONS

By now it is clear that the price of petroleum products is of critical importance to New York in determining electric generating costs, the relative competitiveness of fuels and power in industrial markets, and the level of statewide economic activity. With this background, we turn to an assessment of future oil market conditions and their implications through these linkages.

Our purpose is not to attempt to forecast the path of future oil prices—that path is influenced by unpredictable events—but to indicate some important changes in the structure of the oil market and how these structural changes will affect oil price behavior. In making this assessment, we wish to look beyond the immediate situation, which is one of excess capacity. The pendulum swinging between a loose and a tight oil market may be expected to move toward a situation closely approximating that of the 1970s. When it does, the question arises whether oil prices may be subject to the same degree of instability as experienced in the previous decade.

Our answer is no. Even in the absence of excess supply capacity, the oil market has evolved in important ways that will continue to enhance oil price stability for some time into the future. Two characteristics of past price behavior are unlikely to be repeated in the next several years. First, small interruptions in world oil supply in 1973–74 and 1978–79 caused major adverse reactions by consumers and inventory holders which drove up world oil prices. The effect of increases in demand far outweighed that of reductions in supply in determining the impact on market-clearing prices. Second, once each of the past crises ended, prices did not immediately decline to precrisis levels. OPEC took advantage of the disturbances to establish new, higher official posted prices, causing world oil prices to ratchet upward in two successive jumps (separated, to be sure, by some intervening price erosion in real terms). The 1974 pricing decision did not require stringent production restraints to maintain the new price floor, because demand did not immediately fall. The 1979 decision, in contrast, required tighter production restraints and readjustments in the floor price.

Temporary interruptions in oil supplies are expected to recur from time to time in the future and can be expected to cause increases in oil prices. Nevertheless, compared to earlier experience, the price shocks are expected to be more transitory and less dramatic, even in the absence of the cushion provided by excess capacity. We turn next to the basis for this optimistic view.

Change in OPEC Incentives

The economic interests of OPEC have changed in favor of price stability—not just to maintain current prices but also to avoid price increases. The reason is that the demand for OPEC oil is more elastic than in the 1970s, and oil revenues are likely to fall with a price increase. Similarly, oil revenues are more volatile in response to fluctuations in market supply and demand, so that the price-stabilizing role played by OPEC, acting as the residual supplier, imposes a serious economic burden on and creates a source of tension among member countries. Unlike the previous decade, we would argue, it is no longer in OPEC's collective interest to destabilize the market or to take advantage of instability to raise oil prices.

This shift in OPEC's position is the combined result of sluggish oil demand, of rising oil production from non-OPEC sources, and of increased flexibility by consumers to shift to other fuels as the price of oil rises.[2] OPEC's share of a declining level of noncommunist oil production has declined dramatically—from 67 percent in 1973 to 44 percent in 1983. The reduction in size by itself means that incremental adjustments in production designed to maintain price stability are large in relation to quantity sold and that revenues will fluctuate sharply.

These remarks refer to OPEC as a group; the positions of individual countries will differ from the total. Some members would benefit from a price increase if they bore proportionately less of the burden of supporting the increase; that is, if the quantity sold by those members was not reduced in the same proportion as that sold by the remaining members. Conversely, in today's situation of excess capacity, OPEC as a whole could increase total revenue by permitting the price to fall, while some members of the group would lose. The fact that some members would suffer creates some reluctance to alter the status quo. Added to this is the desire to maintain perceptions of OPEC's market power.

While there are pressures to lower prices in today's slack market, it is important to recognize that the elimination of excess supplies in the future will not by itself reverse these pressures. As long as the demand for OPEC oil remains highly responsive to the price, OPEC has an incentive to avoid higher prices. We expect that this condition will hold beyond that future date when available excess is absorbed and the market tightens.

This does not mean that oil supply crises are a thing of the past but, rather, that OPEC does not have an incentive to engender a crisis for direct economic gain. Oil supplies will continue to be subject to political and military tensions in the Middle East. When the next disruption occurs, oil prices will rise, but our analysis suggests that OPEC will not be inclined to convert a transitory event into a permanent one.

Change in Market Institutions

Like OPEC, the private sector is also less likely to convert minor supply disruptions into major price shocks. The last few years have witnessed significant changes in the way oil transactions are conducted, which promise to enhance price stability during future periods of uncertain supply. We refer to the development of an active spot market in crude oil and porducts, and to the related creation of viable futures markets.[3]

Until very recently, virtually all oil was traded in international markets through long-term contracts or within integrated oil company channels, while the spot market served the limited function of balancing refiners' input and output mixes. The marked increase in the relative importance of spot transactions in all phases of the industry has served to increase the degree of competition in the market and the level of efficiency by which oil is distributed around the world. A measure of the importance of this transition is the corresponding creation of viable futures markets for petroleum where none could survive before.

All segments of the industry have been forced to adjust to these developments. OPEC producers, who once shunned the spot market, have increasingly turned to spot sales as an important outlet and, in the process, have been forced to recognize the competitive link between spot prices and official posted prices. Posted prices may remain stable, but they are discounted in a variety or ways so that effective selling prices track movements in spot prices. Similarly, refiners have been forced into "make-or-buy" decisions based on margins that exist in the spot market, while distributors have to compete with spot and futures prices to make sales in once secure markets. Even individuals outside the industry can "play" in the market by buying and selling in spot and futures markets.

In considering the effect of these institutional developments on oil price behavior, it is important to recognize that they introduce the appearance of price instability at the same time as they create market forces that enhance stability. The appearance of price instability comes from the fact that these prices can fluctuate daily in response to changes in supply and demand conditions, unlike earlier periods when spot prices were firmly anchored to posted prices. Price variability is also a prerequisite for a viable futures market, for otherwise the futures market cannot serve its fundamental purpose of transferring price risk to those willing to bear the burden. Sometimes futures markets are blamed for price variability when in fact these institutions are the result of price fluctuations.

The forces working to stabilize the market generally exert their influence by moderating fluctuations in inventory demand and by countering panic buying in a crisis. These destabilizing activities, and their adverse effect on prices, while unlikely to disappear completely in future crises, will be less detrimental than in the past.

The spot market affects inventory demand because it replaces a traditional need for carrying inventories. Working inventories at refineries and distributors can be smaller because needed supplies may be obtained from, or excess supplies can be sold to, the spot market. Hence, inventory carrying costs can be reduced. Similarly, during a supply crisis, the spot market is less likely to dry up than it has been in the past and will continue to provide an option to affected parties when normal channels are disrupted. Consequently, there is a reduced need to bear the cost of storing oil as insurance against disruptions.

Access to a futures market reinforces these influences. Futures contracts enable those who can least afford to be caught short in a crisis, and those most averse to risk, to cover themselves in a way that is cheaper and easier to arrange than holding inventories of oil. In addition, a futures market provides the opportunity for speculators to gamble on changes in future oil prices without actually dealing in physical stocks of oil. Whether the motive is to hedge or to speculate, transactions involving pieces of paper are easier and cheaper to exchange than barrels of oil.

These developments suggest that future inventory demand will be less influenced by speculation and panic compared to the past. Moreover, wide discrepancies between spot and posted prices are less likely to occur, as there are more opportunities for the market to arbitrage between the two prices, thus removing whatever influences these discrepancies may have on speculative behavior. Finally, it remains to be mentioned that the Strategic Petroleum Reserve is now large enough to exert an influence on the oil market, and the Reagan administration's resolve to use the reserve early in a crisis can have a major calming effect.

Implications for Price Behavior

In short, supply crises in the future may be expected to produce less panic buying and less adverse speculation than in the past. Combined with the hypothesis that OPEC is no longer in a position to exploit supply crises for the purpose of raising official prices to a new plateau, it may be argued that future price shocks will be more transitory and less dramatic than would be expected on the basis of past behavior. To summarize our views regarding future oil price behavior, we expect:

- prices will fluctuate more during normal supply conditions than they have in the past;
- in the event of a disruption in oil supplies, oil prices will oscillate rather than ratchet upward as in the 1970s;
- price shocks will be more moderate in times of crisis because the private sector reactions will be less detrimental to stability;
- the spot market will provide access to oil supplies during future oil crises, so that consumers can get supplies if normal channels are disrupted—but consumers must be prepared to pay market prices;
- the futures market provides insurance to hedge against adverse price changes, to help moderate the loss in income due to a price increase. (However, trading will likely be suspended once a crisis begins, so the insurance must be purchased in advance.)

Implications for New York Utilities

For New York utilities, heavily dependent on oil-fired generating capacity, our message is optimistic. Insofar as the pressures to convert or replace existing oil-fired plants are determined by expectations of rising oil prices (in contrast to past increases), we see additional breathing room beyond the current "softness" in oil markets. In view of the cost of these conversions and the level of excess capacity already available, it might be advisable to delay planned conversions until more favorable conditions prevail. Conversion plans have been stretched out anyway, and from the standpoint of projected fuel costs, the delays need not be burdensome.

While the pressures for oil conversion are less severe, the premium attached to flexible power systems that permit temporary shifts away from oil is likely to increase. With oil prices increasingly subject to business cycles and other market perturbations, together with the ever-present risks of oil supply disruptions, oil is likely to oscillate between periods of favorable and unfavorable competition with other fuels. Consequently, investments in dual-fired generating capacity and in transmission facilities that increase the capability to replace output from oil-fired plants should be given priority. The implications are particularly important for New York City, because of its high dependence on oil and the lack of transmission capacity to transport cheaper power from other areas.

These considerations are less relevant to energy-using industrial firms for whom the advantages of short-term substitutability between oil and gas vastly outweigh the possibility of switching between electricity and fuels. Nonetheless, industry benefits to the extent that flexibility on the part of utilities succeeds in moderating disruption-induced rate increases.

While we claim no insight into specific prospects for New York manufacturing industries, to the extent that the outlook is keyed to energy factors, our analysis of world oil-price developments suggests that the serious dislocations caused by oil price shocks experienced during the last decade may have subsided, at least for some time. If, as a result, economic activity proceeds along a somewhat more even keel, electricity load forecasting may be subject to less uncertainty. Equally important, markets for sales of industrial products will not be subjected to severe fluctuations caused by world oil prices, at least for the rest of this decade. Both nationwide and in New York State, pressures will subside for some form of "industrial policy" to ease the transition problems of firms adjusting to higher energy costs.

For these utilities operating oil-fired generating plants, the need for maintaining large stocks of oil is less critical now than it has been in the past. As we have indicated, access to oil supplies will be less of a problem in future periods of uncertain supply. The problem will be one of temporary fluctuations in the price, but utilities now have a means of hedging against price risks through the futures market. They would be well advised to engage regularly in futures transactions for some fraction of daily requirements over the coming years, to help avoid fluctuations in fuel costs and net income caused by gyrations in the oil market.

Price hedging will help reduce the need for adjustments in electricity rate structures; nevertheless, the flexibility to pass through changes in fuel costs will remain important in the future. As these fluctuations are likely to be frequent—though moderate—it will be important for utilities to have the ability to adjust rate structures quickly. Consequently, fuel adjustment clauses are likely to remain an important feature in rate determination.

Built-in flexibility is also required throughout the power system, beyond the individual plant and utility level. Yet, in many important dimensions, the incentives do not now exist to encourage private investment in flexibility. Often, the reason is that the benefits of flexibility are shared by utilities and by society in general, while the costs must be borne by individual firms. The result is that private investment is insufficient to meet society's needs. This appears to be the situation for investments in transmission capacity required to transport large amounts of power from one region of the country to another.

The same is true for maintaining excess generating capacity around the country that would permit switching away from oil-fired capacity in times of crisis. The costs of excess capacity must be paid by local ratepayers and stockholders; yet the benefits provided by this flexibility accrue to society in general. The benefits are measured in such terms as the balance of payments savings permitted by reducing demand for costly imports and by avoiding unemployment and production losses because of temporary dislocations caused by oil price shocks. In fact, flexibility in the electricity system serves the nation's security interests in the same way as the publicly financed Strategic Petroleum Reserve. Each is a form of insurance that reduces the country's vulnerability to random events in the world. As we see it, such insurance will continue to be critical to the nation's well-being in the future, and methods must be found to cover the costs.

As a final point, what we have said about the evolution of the oil market since the turbulent decade of the 1970s contains an important lesson for future planning. Just as market forces have created incentives for the structural changes that are with us today, we can be sure that the process of change, modified and shaped by unfolding events, will continue into tomorrow. This means, on the one hand, that planners must continually reappraise the situation and not be captive to previous forecasts. It also means that uncertainty about the oil market situation is inevitable and must be included in all long-run investment plans.

NOTES

1. Amory B. Lovins, "Scale, Centralization, and Electrification in Energy Systems," in *Future Strategies for Energy Development* (Oak Ridge, TN: Oak Ridge Associated Universities, 1977), p. 103.

2. These developments are discussed in greater detail in Douglas R. Bohi and William B. Quandt, *Energy Security for the 1980s: Economic and Political Perspectives* (Washington, D.C.: The Brookings Institution, 1984), pp. 2–12.

3. For more details, see Bohi and Quandt, ibid., pp. 12–19.

PART III

ALTERNATIVE ELECTRICITY SUPPLY PLANS AND THE ROLE OF EVOLVING TECHNOLOGIES ——————————

Although the different forecasts of long-term electricity demand presented in Part II are surprisingly similar, given the substantial variation in underlying assumptions regarding economic growth and relative fuel prices, nevertheless, all of the authors emphasize how consistently historic load growth projections have been wrong. The authors in this section on alternative electricity supply plans seem highly conscious of that lesson, because they all emphasize the need to incorporate diversity of supply in their plans. This hedging is in part an attempt to account for future uncertainties about the level of future demand, the potential for supply disruption, and the relative price of primary fuel supplies and of construction costs.

The other thread connecting the chapters in Part III is an awareness of the public's emerging ambivalance toward large-scale technological solutions of society's needs, the so-called technological fix. This appears vividly in the debate raised by Percival regarding the future role of conservation and of small-scale decentralized generation sources like wind, hydroelectric, and co-generation in offsetting or meeting future electricity demand. Utility planners and public agencies responsible for ensuring that adequate means are available to meet future electricity demand, while acknowledging the importance of conservation and co-generation, seem unwilling to rely too heavily on these alternative sources, in part because they are decentralized, which leads government and utility planners to feel less certain about their ability to control the development of these fragmented sources. Conversely, proponents of these alternative sources point to the problems encountered in the previous decade with bringing large-scale central station plants on-line at reasonable costs, and they emphasize that in an era of uncertainty, building a variety of many small units may make more sense than building a few large facilities.

What both sides of this debate agree on is that currently in New York—and throughout much of the nation—there is ample generation capacity to meet foreseeable electricity demands over the next 10–15 years. What is disquieting to utility and government agency planners like Davis, Stuzin, and Hiney is that a substantial fraction of the available capacity is old, inefficient, and/or burns expensive imported oil, and that the gestation period from inception to operation of a new central station unit is ten years or longer. Thus if additional capacity will be required in New York before the year 2000, utility planners ought to begin planning almost immediately. Meanwhile advocates of conservation and alternative decentralized generation sources point to the much shorter lead time needed for completion of their proposed supply-side alternatives; they argue that the proper incentives (or lack of regulatory disincentives) could encourage the utilities to turn to these unconventional solutions to fulfill their supply needs.

Weinberg emphasizes that many of the "new" solutions to potential future supply needs embody old, well-established technologies. What has changed in the past decade to heighten interest in conservation and co-generation are the relative prices of fuels and equipment, and regulation. Debate over how and to what extent the existing institutional/regulatory framework is biased for or against large central station units will be addressed in greater depth in Part VII. Part III indicates that conservation, hydroelectric power, and co-generation technologies are established and viable alternatives for meeting a portion of New York's needs. The final two authors in this section (Weinberg and Huband) also leap beyond currently available technologies to survey a range of emerging technologies that may affect the economic viability of both central station and decentralized sources of supply beyond the year 2000.

9

New York's Electricity Supply: Present Capacity and Future Needs

William E. Davis

Uncertainty reigns supreme in electricity forecasting and planning. New York State has long been in the forefront of efforts to plan rationally for a better energy future, and we have seen significant improvement in the state's overall energy posture; however, we still have a long way to go before we should consider our energy problems under control. In general, society has not been nearly as successful in implementing energy policies over the past few years as it has been in planning them— especially in the electricity sector. Our overall energy situation in New York State is somewhat like a good news/bad news joke. The good news is that we have improved on almost all fronts since the mid 1970s. The bad news is that because we started out in such a poor situation, we still have a long way to go.

Over the period 1973-83, total primary energy use in New York State declined by approximately 21 percent. Over the same period, total petroleum consumption declined by 38 percent. Another, and perhaps better, measure of the improvement in our energy situation over this period shows our energy consumption per unit of gross state product declining by 25 percent, indicating that the energy efficiency of the state's economy improved substantially during this time. Moreover, our energy use per capita declined by 22 percent over this period, which also suggests increased energy efficiency across the state.

On the bad news side, as I indicated, we still have a long way to go before we should consider our energy problems solved. We still import 90 percent of our primary energy. Our energy bill in New York State is now over $26 billion a year and almost $16 billion leaves the state to pay for our energy imports. The result, in general, is that our energy prices are too high and our energy supplies are still too insecure.

Regarding the state's electricity situation, again we have a good news/bad news situation. The good news is that we have plenty of capacity. Our current reserve margin is over 40 percent. Our mix has also improved considerably over the last few years. Petroleum as a percentage of our primary fuels used to generate electricity is down 17 percent since 1978.

On the bad news side, the excess capacity that we have in New York State is generally old, inefficient, and expensive oil-fired generation. Also on the bad news side, one of our primary problems is that our electricity costs too much, at least in some parts of the state. The overall average cost of electricity in New York State is 40 percent higher than the average cost of electricity in the rest of the nation.

In Charles Guinn's chapter on energy forecasts, Figure 5.1 illustrates how electricity fits into the overall energy situation. The electricity sector consumes about 39 percent of the total primary energy used in New York and generates about 15 percent of the end-use energy. Figure 5.1 also indicates the state's overall heavy dependence on petroleum products to generate electricity. Of the electricity that is generated in New York State the largest portion, 44 percent, goes to commercial customers, 30 percent is used to supply residential needs, with 24 percent being used to supply industrial needs. The remaining 2 percent is used in transportation.

Figure 9.1 illustrates how New York State compares with the United States in regard to fuels used to generate electricity. New York State is more heavily dependent on pertroleum, uses more hydroelectric power to generate electricity, imports more electricity, and is much less dependent on coal than is the U.S. as a whole. Compared with the country as a whole, New York State's proportions of nuclear and natural gas-generated electricity are approximately the same.

A simple summary of New York's situation is as follows: our total installed capacity, including the recent addition of the Somerset plant, is 31,814 megawatts (MW); peak load in New York State for the summer of 1984 was 21,971 MW; and the resulting statewide reserve margin was approximately 45 percent.

Another way of viewing these statistics is that New York currently has about 5,000 MW of excess capacity above the 22 percent reserve margin required to ensure reliable service. Using projections by the State Energy Office (SEO), only 3,000 MW of new load will be added over the next sixteen years. At the same time, we project about 1,600 MW of retirements to occur over the same period. Therefore, assuming these forecasts are correct, New York will still have more than the required 22 percent reserve margin in 1999, without adding any of the currently planned units or those under construction: Shoreham and Nine Mile Point Two nuclear plants, or the Prattsville pumped storage facility. This projec-

FIGURE 9.1 Primary Consumption by Electric Utilities, 1983

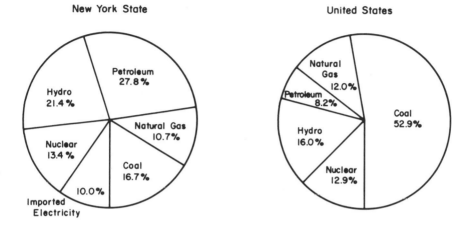

lion also does not consider any additional facilities that use renewable sources. Of course, if our view of the future demand for electricity is too low, New York would need some of the aforementioned capacity to maintain minimum reliability standards.

This simplified view of what is required to keep the lights on in New York State has, however, some drawbacks, If no new capacity is added over the next fifteen years, the state's utilities may be putting all of that old, expensive-to-operate, oil-fired capacity back into operation more frequently in order to meet demand, with the result being higher costs, less competitive rates, and greater adverse environmental impacts.

The challenge for New York State electricity planners over the next decades is not associated so much with keeping the lights on as it is with providing economical electricity that is produced in environmentally acceptable ways. While some upstate electric rates are competitive nationally, the electric rates of downstate investor-owned utilities are among the highest in the country and in some cases are more than double those of upstate utilities. To compound this underlying economic disparity, some of the recent efforts to provide new economical sources of electricity in New York, namely the Shoreham and Nine Mile Point Two nuclear projects, have produced two of the most expensive generating plants in the country, and the bulk of their rate impact has yet to be seen in customers' bills.

These problems have arisen even though New York State has long been at the forefront of efforts to plan for electric energy in a comprehensive and integrated manner. As an example, the State Energy Master Plan (SEMP) process has incorporated, in a systematic and structured way, consideration of the impact of energy efficiency improvements on demand for all energy forms, including electricity; consideration of the development of alternative energy technologies; and consideration of the need for new conventional generating facilities. The objective of this process, and of the plans and recommendations it has generated, is to ensure adequate, reasonably priced, reliable, and secure electricity for New York consumers in a manner that balances environmental, economic, and financial considerations.

The most recent electricity supply plan resulting from this planning process, which was contained in the draft SEMP 3:

- assumed completion and operation of those plants already under construction (Shoreham, Nine Mile Point Two, Somerset);
- recommended licensing and construction of a 1,000-MW pumped storage hydroelectric facility;
- recommended oil-to-coal conversions of seven generating plants totaling over 2,500 MW;
- projected development, assuming implementation of certain recommended incentives, of almost 3,000 MW of alternative electricity generation sources, such as small hydro, co-generation, solid waste, and wind;
- projected importation of as much as 24 billion kWh of electricity from Canada.

The effect of implementing this plan on the primary energy used to generate electricity is depicted in Fig. 9.2. As indicated, New York's dependence upon petroleum to generate electricity would be reduced from 28 to 14 percent. The nuclear share would increase from 12 to 21 percent. Imports would increase from 8 to 12 percent, and renewables and small power production could become a significant fraction (10 percent) of the total mix by 1999.

Substantial benefits are projected to result from implementation of this plan, including $8.5 million in net energy cost savings; $4 billion in increased total earnings; and 227,000 additional employee-years of employment over the planning period. On an average annual basis, these last two benefits translate into approximately $240 million a year in earnings and creation or retention of 13,000 jobs.

However, the draft SEMP 3 was never finalized. The legislative authority for the planning process expired on January 1, 1984, prior to the finalization of the SEMP 3 process, and prior to approval of the plan by the State Energy Planning Board. Perhaps more important, it is apparent that some of the major assumptions regarding electricity supply that underlie the draft SEMP 3 are questionable. As examples:

FIGURE 9.2 Primary Energy Used to Generate Electricity in New York State, 1982 and 1999

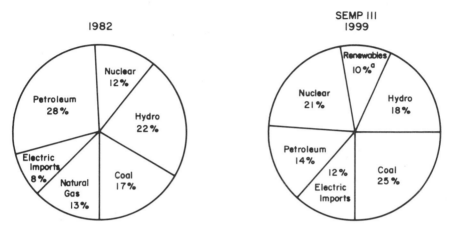

ᵃIncludes additional small hydropower generation.

- The lack of an acceptable emergency evacuation plan has resulted in considerable uncertainty over whether Shoreham will ever operate.
- It is doubtful that the Prattsville pumped storage project will go forward, unless changes can be made to ensure that state environmental standards can be met.
- Many of the future coal-conversion projects are bogged down in regulatory review proceedings or legal challenges.
- Although progress is steady and encouraging on may small power projects across the state, uncertainty with regard to future oil prices, tax credits, and other incentives calls into question whether or not it is realistic to project that 3,000 MW can be installed in New York over the next sixteen years.
- Falling oil prices combined with increased economic activity in New York State have resulted in increasing uncertainty regarding future load growth projections.

While Somerset has been completed and progress has been encouraging on installing conservation devices, siting transmission lines, arranging for Canadian imports, and even on constructing Nine Mile Point Two, it is clear that more planning, not less, is called for. During the most recent session, the New York State legislature and the governor's office came very close to reaching agreement on a new planning process.

Looking toward the future, electricity planning in New York State will in my opinion require the following:

- legislation to renew the comprehensive planning process;
- greater emphasis on cost-effective conservation;
- further upgrading of the transmission system to give us greater flexibility to move economic power to where it is most needed;
- more emphasis on small power production;
- a concerted effort toward overhauling exisiting older facilities in order to prolong their economic life.

In my view, these are the most appropriate strategies for meeting the challenge of providing more economically attractive and environmentally acceptable electric energy to fuel both households and New York's growing state economy.

10

Existing Electric Generating Capacity in New York State ─────

Lester M. Stuzin

Is there a crisis today or will there be in the near future of the availabity of electricity in New York State? That is the question asked by the editors of this volume. As far as today is concerned, the answer is no. In the near future, through 1990, the answer is also no. After that, depending on load growth, the answer is unclear.

Forecasting in the area of load growth has had a very poor track record over the last fifteen years because of exogenous events. During the past two years, load growth has been higher than anticipated throughout the nation. Any significant change upward in the State Energy Office (SEO) projection of less than 1 percent a year could necessitate a reevaluation of the required capacity situation. The New York Power Pool (NYPP) projected a 1.4 percent load growth in the same two-year period.

At present, there is more than adequate electric capacity available to meet the needs of New York State's consumers. In addition to the existing electric-generating capacity in New York, there are major interconnections with Canada, New England, and the Pennsylvania/New Jersey/Maryland pool.

However, I must add that electric generation in New York State is not economical. Our ratepayers have had to pay the penalty for burning too much expensive oil. For example, in 1983, 27 percent of the electric energy was produced by oil-fired generation, compared with 7 percent nationwide. Considering that 60 percent of the installed capacity in New York is oil-fired, this low usage rate reflects a major reduction in oil dependence that has been made possible principally by increasing electricity purchases from Canada and by increasing the use of natural gas for electric generation.

It had been hoped that the reduction in oil usage could have been even greater by converting oil-fired boilers, used to produce steam for driving turbine generators, to burn coal. The New York utilities have not been able to accomplish any oil-to-coal conversions since the 1973 oil embargo, even though conversion has been a state and federal policy for many years. The primary obstacles have been environmental concerns and uncertainty over the potential costs of achieving satisfactory environmental compliance.

The New York State Public Service Commission (PSC), the New York State Energy Office, the federal government, and the utilities all agree that coal conversion is essential to reduce New York's excessive dependence on foreign oil for electric power generation. Such a reduction would reduce costs of electricity to ratepayers and make the system more reliable by diversifying energy sources.

Some of the utilities rely more heavily on oil than others. For example, all of the Long Island Lighting Company's (LILCO) 3,721 megawatts (MW) and 8,556 MW of Consolidated Edison's (ConEd) 9,405 MW are generated by burning petroleum. An oil embargo could have a major impact on these two utilities since the transmission system is not adequate to bring in all their energy needs from outside the New York metropolitan area.

During the five-year period 1980–84, at the time of the NYPP system peak, the installed generating capacity reserve ranged from 38 to 42 percent, well above the installed reserve requirement for the pool, which is currently 22 percent. I must add that the reserve margin and surplus to a great extent consist of old, inefficient oil-fired units and combustion turbines. To put that fact in perspective, consider that in the years 1980–84, when the installed reserve margins ranged from 8,300 to 9,100 MW, the NYPP had an oil-fired steam capacity of over 14,000 MW, and combustion turbines of about 3,500 MW. That represents about 60 percent of the installed capacity in the state—and that oil-fired capacity consumes very expensive low sulfur oil. By comparison, about 20 percent of generating capacity nationwide is oil-fired.

In viewing electric-generating capacity in New York State, it is also important that actual or operating reserve margins be considered, not merely the total number of megawatts installed. The total installed capacity does not reflect the day-to-day operation of the system: some units are always unavailable because of maintenance, forced outages, and more recently, deratings as the average age of units increases. Nonetheless, in the past operating reserves have been adequate because of the availability of interconnections. Our strongest resource to help us meet unforeseen contingencies and forced outages are our transmission ties to Canada. Because New York State is essentially a summer-peaking area and the Canadians are winter peaking, we can generally count on them for several thousand megawatts during the summer months.

During the 1980–84 period, unavailable capacity during the summer peak (from mid-June through mid-September) ranged from 7,200 to 8,200 MW. This range represents 24 to 27 percent of the total installed capacity. The unavailable capacity results from several factors, including the large number of oil-fired units (which are less well maintained because of their high cost of generation) that make up the NYPP; the "normal" anticipated forced outages of units owing to equipment failures; and perhaps to some extent, the declining reliability of units because of the increasing average age of capacity in New York State.

Should this high amount of unavailable capacity be a concern to electric users in New York State at this time? I do not believe so. With the large installed reserve and the use of the interconnections, especially to Canada, it has not been a problem in the past decade.

In addition, a portion of the available capacity that is down for maintenance could be made available if different maintenance schedules were decided on by the NYPP. However, at present there is no need to schedule maintenance to ensure the availability of adequate capacity; therefore economics are the primary consideration in scheduling crews to complete repairs. As an example, if a high-cost oil-fired unit were down , a New York utility would not schedule overtime or bring in an outside contractor to repair the unit. The repairs would be performed during regular shift hours.

Electric users should be concerned about the age of New York's electric generating system. Fully 25 percent of the NYPP fossil fuel generation is over thirty years old. A NYPP study for the years 1974–78 showed that the forced outage rates for fossil fuel-fired units increased notably after thirty-five years.

The question of availability and efficiency of older units is being addressed by the PSC and the utilities. The PSC staff, with the assistance of consultants, has undertaken an evaluation of power plant performance for the seven largest electric companies to determine how the availability and efficiency of the entire system might be increased. The thrust of the study has been to encourage the utilities to analyze equipment failure and forced outages in order to determine underlying causes. With this information, systematic improvements in maintenance procedures and records, and increases in the availability and lifespan of equipment may be possible.

As time passes, the importance of extending the life of these units will increase. An example of an approach that is being taken is the work the New York State Electric and Gas Company has started on its coal-burning units. The company estimates a cost of $185–$255/kW to rehabilitate two of its stations in order to extend their lives by 20 years. These units were originally constructed during 1943–1950. By comparison, the construction of new coal-fired capacity costs $1,000–$1,500/kW, with an estimated 30–35 year lifespan.

In summary, the current generating system appears adequate to supply today's electric loads and the loads for the near future, but even moderate increases in the rate of load growth could result in shortages by 1999. Furthermore, much of this existing capacity is oil-fired and thus uneconomical. Future disruptions in oil supplies from the Mideast will only exacerbate this situation.

11

Importing Canadian Electricity——
Robert A. Hiney

This chapter will focus on a single but important part of New York State's electric energy plan: Canadian electricity purchases and related transmission lines.

The New York Power Authority (NYPA) has been a major purchaser of Quebec's electricity since energizing the 765-kilovolt (kV) interconnection in 1978. In 1983, of the 39 billion kilowatt-hours (kWh) supplied by the NYPA, nearly 9 billion were purchased from Hydro Quebec. Altogether, New York State's electric utilities purchased nearly 20 billion kWh from Canada: 10 billion kWh from Quebec and 9.9 billion kWh from Ontario. That amount was about 16 percent of the electricity used in New York State. It was also more than one half of the total imports of Canadian electricity into the United States.

Most of the purchases were not for firm power [1] but, rather, for fuel replacement or economy energy[2] used to displace fossil fuel—coal, gas, and oil. Some critics say that New York has become dependent on Canadian power; however, that is not the case. The Canadian purchases can be replaced with additional production from existing generating plants in New York, but at higher cost. The NYPA and the other utilities are simply using the cheapest available sources to the maximum practical extent.

The electricity from Ontario Hydro is produced mainly at coal-fired plants. It is available for export as a result of reduced load growth in Ontario and because several nuclear plants recently completed, or under construction, will help meet the province's domestic needs. The electricity from Quebec is all hydroelectric. Hydro Quebec's large program to develop its hydro resources—notably, the James Bay Project, which has an initial capacity that is four times that of the Niagara hydroelectric project—coupled with declining projections of future load growth have made large surpluses available for export to the United States.

Canadian hydro-power is often called "cheap Canadian hydro-power," and it is cheaper than the power it replaces. However, it is often confused with the NYPA's own hydroelectric power produced at the Niagara and St. Lawrence River plants, which were built more than 20 years ago. *That* power is not just cheap—it is about the least expensive power in the country. At less than one-half cent/kWh, it costs about one-sixth as much as the Quebec hydroelectricity purchased in 1984. But the Quebec electricity in turn provides substantial savings, averaging about 60 percent of the cost of generation in an efficient oil-fired plant such as the NYPA's Poletti unit in New York City.

A drawback to the Canadian power is its point of origin. On the one hand, New York State shares a border with both Quebec and Ontario. This common border permits construction of direct transmission ties to each province. These are advantageous. On the other hand, the source is far from the market. Hydro Quebec's James Bay project in northern Quebec is several hundred miles from the border. Unfortunately, the principal market for the purchased Canadian power is even farther south—the bulk of the high-cost oil-fired generating units are located in the lower Hudson Valley, New York City, and Long Island. To benefit the state most, the Canadian power must displace New York's most expensive generating plants. Thus, it should flow to southeastern New York.

The NYPA's 765-kV transmission line takes power from Quebec and Ontario to Utica, New York, 150 miles south of the border. This line was built in connection with an 800-megawatt (MW) firm power contract with Hydro Quebec. Fortunately, the NYPA was given approval by the Public Service Commission (PSC) to construct a line able to transmit much greater quantitites of power. The NYPA has been importing more than is covered by that contract; moreover, electric substation equipment recently installed by the NYPA and Hydro Quebec at each end will increase this line's capability to about 2,300 MW. That is almost three times the capacity involved in the NYPA's initial contract.

However, there is a bottleneck south of Utica. To deliver the additional power from Utica to southeastern New York, the NYPA is seeking approval from the PSC[3] to build the Marcy South 345-kV transmission project. This project will upgrade the existing network primarily by implementing double-circuit transmission lines that link five existing substations and a new one at the southern end of these lines. These improvements will provide a connection into Consolidated Edison's existing lines near Fishkill in Dutchess County. The Marcy South plan will increase the capability of the transmission network to move power across New York State by about 2,500 MW, an increase of 65 percent.

Often overlooked is another feature of transmission lines—their contribution to the reliability of our electric service. Transmission is the best resource we have for dealing with the unexpected. Whether it is a coal strike, an oil embargo, or a shutdown of a major generating plant, power lines, or substation, the transmission network with double-circuit plan will be more lightly loaded and better able than a lesser reinforcement, such as single circuit or lower voltage lines, to withstand a major emergency without a blackout.

I would like to conclude with some thoughts on the prospects for future purchases of Canadian electricity. First, the market in New York will remain large. By the year 2000, oil consumption for electricity is projected to be larger than it was in 1983; thus ample opportunity to displace more oil will exist. How much Canadian electricity is purchased will depend on its availability, the cost of alternative sources, and, of course, the ability to get it to the market.

One thing is sure: Canadian energy purchases can continue to be a significant element in New York's plans long after the ten-year period (1987–1997) analyzed to justify Marcy South. The interconnection exists to allow mutual assistance during emergencies as well as economy energy transactions. Hydro Quebec has 30,000 MW of undeveloped water resources and is quite willing to speed up its hydroelectric development program to ensure large exports to U.S. markets well into the next century; this will likely depend on firm commitments by U.S. buyers. That prospect, along with the alternative future supply options, will have to be carefully evaluated.

NOTES

1. Commitments to meet a specified amount of the purchasing utility's capacity needs.

2. Sales of kWh, when excess energy is available from the seller that is economical for the buyer.

3. The PSC subsequently approved this project.

12

Conservation and Renewable Energy Sources as Supply Alternatives for New York's Electric Utilities————

Robert V. Percival

The environment in which New York's electric utilities operate has changed dramatically during the last decade. Like many others, utilities in New York in the early 1970s launched ambitious plans to construct additional power plants based on projections of rapid growth in demand for electricity. These plans later had to be sharply scaled back when the projected demand failed to materialize following the Arab oil embargo. But at no time did the utilities question whether their customers' energy needs could be satisfied more economically through investments in alternatives to central station power plants. Their utilities' failure to consider alternatives is costing New York consumers dearly today; however, there is hope that New York's utilities eventually will pursue conservation and renewable energy investments as more economical alternatives to construction of central station power plants.

THE CONSERVATION INVESTMENT CONCEPT

When utilities operated in an environment of declining marginal costs, expanding demand for electricity was accompanied by reductions in the real price of electricity. Early skirmishes between environmentalists and utilities focused not on the economics of electricity generation but, rather, on the environmental impacts of power plants and on the question of how many additional power plants were needed to meet demand growth. The debate often concentrated on the accuracy of forecasts of future demand growth. Few questioned the implicit assumption that the only way to satisfy additional demand was to build more central station power plants.

During the decade of the 1970s, utilities ceased to operate in an environment of declining marginal costs. For many utilities construction and operation of new central station generating capacity became more expensive than utilization of exisiting generation sources. The increasing costs of new power plant construction and concern over environment degradation spurred a search for alternatives.

In the late 1970s the Environmental Defense Fund (EDF) helped pioneer the notion that electrical utilities could benefit themselves and their customers by investing directly in end-use conservation devices. The EDF's argument was quite different from what utilities were accustomed to encountering. It did not challenge utility projections of future demand growth; it focused instead on the question of how utilities could satisfy most economically whatever customer demand they forecasted.

The EDF maintained that strictly as an economic proposition central station power plants were not the best investment alternative for utilities. The EDF demonstrated that California's two largest utilities could meet all their projected demand growth at substantially less cost and with substantially less financial risk if they canceled plans to build additional coal and nuclear power plants and instead invested in conservation hardware and small-scale renewable energy sources.

After a vigorous skirmish, California's two largest utilities abandoned their plans to build additional central station power plants before the end of the century. Pacific Gas and Electric and Southern California Edison announced that they would replace these plants in their supply with investments in end-use efficiency improvements; geothermal, wind, and solar energy; co-generation; and conservation voltage regulation.[1]

THE EDF ALTERNATIVE PLAN FOR NEW YORK UTILITIES

The EDF turned its attention eastward in 1981. Five New York utilities were struggling to build a 1,080–megawatt (MW) nuclear power plant, Nine Mile Point Two, which had been plagued with massive cost overruns. In September 1981, the New York Public Service Commission (PSC) ordered a special hearing to consider alternatives to completion of the Nine Mile Point Two project. The EDF presented a comprehensive plan for replacing Nine Mile Point Two with investments in conservation and small-scale renewable energy sources.

When the Nine Mile Point Two project was launched in 1971, its sponsors estimated that it would be completed in 1977 at a cost of $370 million. By 1981, the utilities' cost estimate had risen to $3.7 billion for completion in October 1986. Figure 12.1 shows the rapid escalation that had occurred in the utilities' cost estimates for Nine Mile Point Two.

**FIGURE 12.1   Estimated Final Cost of Nine Mile
Point Two by Year of Estimate**

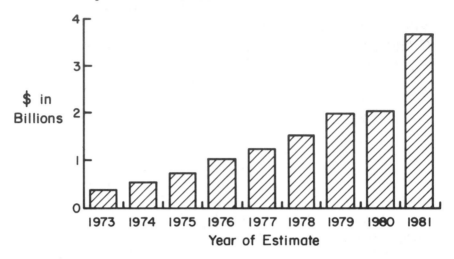

Source: Co-tenants' Response to CPB Interrogatory No. 2, October, 1981, in New York
Public Service Commission Case No. 28059, Proceeding to Inquire into the Financial and Economic Cost Implications of Constructing the Nine Mile Point Unit 2 Nuclear Station; New
York State Power Pool, Section 149-b filings from 1973 through 1978, and Section 5-112 filings from 1979 to 1981.

By the time of the PSC hearing, more than $1 billion had been sunk
into the Nine Mile Point Two project. Despite this enormous sunk cost,
EDF's economists, using a sophisticated computer model to simulate utility investment decisions, demonsrated that the utilities would be far better off financially if they abandoned the plant and pursued alternative investments instead.[2]

The EDF alternative plan, whose components are outlined in Fig.
12.2, involved utility investments in residential and commercial sector
end-use efficiency improvements, co-generation, conservation voltage
regulation, and small hydroelectric projects at existing dam sites. The end-
use efficiency investments, which constituted more than 55 percent of
the plan, included in the residential sector water heater insulation,
fluorescent lighting, low-flow showerheads, and energy efficient refrigerators and air conditioners; and in the commercial sector fluorescent lamps
and energy-efficient ballasts as well as more efficient heating, ventilation,
and air conditioning systems.

The EDF maintained that these investments could provide the energy
and capacity equivalent to Nine Mile Point Two with equal (or greater)
reliability during the same period of time at a cost 17 percent less than
the cost of completion of the plant. These cost savings were particularly

FIGURE 12.2   The EDF Alternative

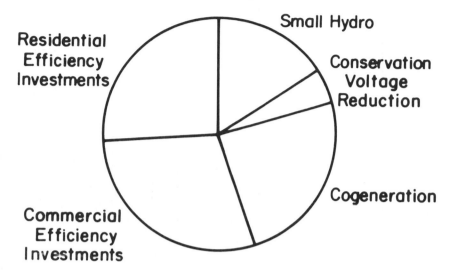

Source: Environmental Defense Fund, *A New Alternative to Completing Nine Mile Point Unit 2 Nuclear Station: Economic and Technical Analysis (November, 1981), p. II-6.*

significant because they were calculated on the assumption that ratepayers bore the full sunk costs of the canceled plant.

The EDF's alternative plan represented an entirely new approach for New York's investor-owned utilities. The EDF's proposal was based on the notion that utilities should treat conservation investments as a source of energy services on an equal footing with traditional supply alternatives. By offering financial incentives to stimulate conservation investments or by directly supplying their customers with conservation hardware, utilities can avoid constructing additional generating capacity and can reap financial benefits for themselves and their ratepayers. To provide an incentive for utilities to pursue conservation investments, the EDF proposed that utilities be permitted to earn a rate of return on these investments similar to what they earn on investments in conventional supply alternatives.

THE UTILITIES' RESPONSE TO THE EDF PLAN

None of the utilities involved in the Nine Mile Point Two project had ever considered direct utility investments in conservation and small scattered renewable sources of supply as an alternative to construction of the plant. Because they could not dispute the notion that such alternatives would be cheaper than completion of a multibillion dollar nuclear power plant, they sought to avoid any direct cost comparison between the two. Instead of arguing comparative economics, they sought to attack the concept of utility investment in conservation, even though it already had been embraced by utilities in many other states.

Legal and Philosophical Objections

The utilities argued that direct utility investments in conservation would pose difficult legal problems. Yet they refused to be specific about what legal obstacles they foresaw, hoping merely to create sufficient doubt to diminish the appeal of the EDF alternative. One of the "legal" arguments articulated during the Nine Mile Point Two proceeding was that conservation investments would require approval from state regulatory authorities—hardly an obstacle since those authorities were the very body conducting the inquiry into alternatives to Nine Miles Point Two.

In subsequent PSC proceedings focusing on the EDF alternative,[3] the utilities were finally forced to articulate their legal objections to direct utility investment in conservation. Aside from the utilities' claim that the PSC had no authority to order them to invest in conservation (even if such investments were the most economical means of providing service to their customers), the principal legal objection they raised was their fear of running afoul of the antitrust laws. The administrative law judge who presided over the conservation proceeding agreed with the EDF that the antitrust laws did not bar utility rebate programs and that the utilities would be insulated from antitrust liability in any event by the "state action" exemption from the antitrust laws. The judge ruled that there are no legal barriers under federal or state law to investment by utilities in conservation in New York.

The more basic objection of the utilities to the EDF plan was philosophical rather than legal. Most of the New York utilities believe that their business is to sell electricity, not conservation.[4] They view conservation with as much enthusiasm as a cigarette manufacturer has for an antismoking campaign. For public relations purposes they favor informational advertising by utilities to promote conservation, but they are fearful that too many customers may begin to heed the message.[5] Objections based on what utilities conceive to be their traditional business role are difficult to justify, however, in the face of evidence demonstrating that it is far more costly to utilities and their ratepayers if utility investment choices are limited to traditional alternatives.

Uncertainty of Customer Response

Although other utilities throughout the country had implemented conservation investment programs successfully, the New York utilities argued that such programs would not work in New York because their customers would not respond to conservation incentives. Because none of the New York utilities had ever considered such programs, they had little basis for claims that New York customers would react differently. It was not until they were required to experiment with such programs during the PSC's conservation proceeding that they generated data specific to their service territories.

The results of their experiments were dramatic. Niagara Mohawk, which conducted the most extensive conservation incentive experiment of the seven investor-owned utilities, offered several different rebate and installation programs. Figure 12.3 shows the customer response to a Niagara Mohawk program offering rebates to customers who purchased water heater wraps. As the chart indicates, when the utility offered a full rebate to customers who purchased water heater wraps, in less than one month approximately 13 percent of the customers in the sample purchased water heater wraps and applied to the utility for reimbursement.[6] This represents an extraordinary response considering the short duration of the experiment and the fact that customers had to go out and locate the conservation device and mail a form into the utility in order to be reimbursed.

Niagara Mohawk also offered a rebate program to customers who purchased energy efficient fluorescent light bulbs. Although the utilities

FIGURE 12.3 Customer Response to Niagara Mohawk Offer of Rebates for Purchasing Water Heater Insulation

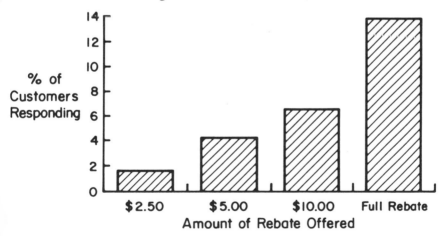

Source: Exhibit 100 in New York Public Service Commission Case No. 28223, Proceeding to Inquire into the Benefits to Ratepayers and Utilities from Implementation of Conservation Programs that Will Reduce Electric Use.

went to great lengths to attempt to establish that few customers could use such lights,[7] the Niagara Mohawk rebate program was a tremendous success, as Fig. 12.4 demonstrates.

Niagara Mohawk also experimented with a direct mail program to distribute low-flow showerheads to their customers. This program did not require the customers to locate and purchase the showerheads themselves. It simply required customers to return a coupon to Niagara Mohawk requesting the showerhead. Niagara Mohawk experimented with different incentive levels in the showerhead program. The results are shown in Fig. 12.5. Of the customers offered the showerheads for $7.00, 10 percent accepted the offer. It is interesting to note that because the showerheads only cost Niagara Mohawk $5.08 each, the company was able to recover more than the cost of the conservation device from these customers. As the level of incentive increased, so too did the customer response. Of the customers offered free showerheads, more than 44 percent accepted the offer in less than one month.

Niagara Mohawk also offered certain customers the option of having the light bulbs, showerheads, or water heater wraps installed for free. The utility hired a contractor to provide this service to a group of Niagara Mohawk customers. As Fig. 12.6 demonstrates, in less than one month, 50.6 percent of Niagara Mohawk customers who were offered free installation of these devices accepted the offer. This demonstrates

FIGURE 12.4   Response to Niagara Mohawk Offer of Rebates for Purchasing Energy Efficient Lighting

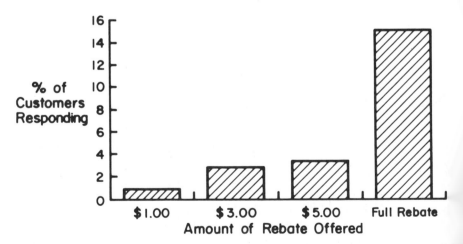

Source: Exhibit 100 in New York Public Service Commission Case No. 28223, Proceeding to Inquire into the Benefits to Ratepayers and Utilities from Implementation of Conservation Programs that Will Reduce Electric Use.

FIGURE 12.5 Response to Niagara Mohawk Program of Direct Mail Distribution of Low-Flow Showerheads

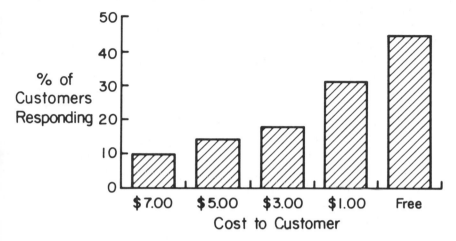

Source: Exhibit 100 in New York Public Service Commission Case No. 28223, Proceeding to Inquire into the Benefits to Ratepayers and Utilities from Implementation of Conservation Programs that Will Reduce Electric Use.

FIGURE 12.6 Response to Niagara Mohawk's Free Distribution or Free Installation Offers, by Device

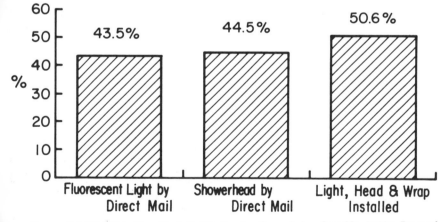

Source: Exhibit 100 in New York Public Service Commission Case No. 28223, Proceeding to Inquire into the Benefits to Ratepayers and Utilities from Implementation of Conservation Programs that Will Reduce Electric Use.

dramatically that customers will respond favorably to conservation incentives provided by the utility.

Unfortunately, Niagara Mohawk's lawyers chose to interpret these splendid results in the most unfavorable light. Although their pilot programs had given customers less than a month to respond, they characterized the results as proof that no more than half their customers ultimately would respond to a full-scale program of conservation incentives. This interpretation illustrates how determined the utilities have been to oppose direct utility investments in conservation.

During the Nine Mile Point Two proceeding, the utilities also argued that the EDF had overestimated the energy savings from certain conservation investments because customers had already undertaken conservation measures not accounted for by the EDF. For example, they argued that the EDF had erred in calculating energy savings from water heater wraps because it had based its calculations on the assumption that electric water heaters in New York were set at an average temperature of 140 degrees Fahrenheit. When questioned as to whether they had any actual data on the average temperature of water heaters in their service territories that would contradict the EDF's assumption, the utility witnesses conceded that they did not but maintained that it would be irresponsible to cancel a nuclear power plant in the absence of such information. The EDF argued that it would be irresponsible to decide to proceed with a multibillion dollar investment simply because the temperature at which their customers set their water heaters was unknown.

During the conservation proceeding that followed the Nine Mile Point Two hearings, the utilities conducted a detailed survey of end-uses of electricity in the residential sector of their service territories. This statewide survey, conducted in 1983, found that the mean temperature at which New York customers set their electric water heaters was 140 degrees Fahrenheit.

The "Do Both" Response

The utilities strove mightily to avoid comparing the economics of conservation investments with the cost of completing Nine Mile Point Two. Their principal argument in response to the EDF alternative plan became known as the "do both" argument. The utilities maintained that conservation investments should be considered a complement to rather than a substitute for Nine Mile Point Two. Because New York was so heavily dependent on oil-fired generation, the utilities argued, any alternative to oil-fired generation made economic sense. Thus, they argued that Nine Mile Point Two should be completed, even if it were more expensive than the implementation of conservation efforts, because it would produce net savings when compared to expensive oil-fired generation. This "do both" argument maintained that New York ratepayers would be better off if the utilities invested in both Nine Mile Point Two *and* full-scale conservation programs.

The utilities' argument ignored the question of which investment provided the highest return. Because the utilities had no intention of investing in conservation, they were essentially arguing that even if Nine Mile Point Two were an inferior investment to the EDF plan it should be pursued instead because it offered some improvement over the status quo. The utility argument also failed to take into account differences in relative risks between continuing the Nine Mile Point Two project and pursuing the EDF plan.

Experiences since the PSC decision approving completion of Nine Mile Point Two shows the dangerous consequences of accepting the "do both" argument. The economic losses from the Nine Mile Point Two project have increased, while pursuit of more economical conservation alternatives has been delayed as a result of the utilities' involvement in Nine Mile Point Two.

At the time of the PSC hearing, the utilities estimated that Nine Mile Point Two would be completed at a total cost of $3.7 billion. The PSC staff maintained that this estimate was unrealistically low and that the project would cost $4.9 billion for a 1987 completion date. (The EDF employed the PSC staff estimate in its comparative analysis.) Costs of the project have now soared far above even the PSC staff's estimate.

Figure 12.7 shows the magnitude of the actual cost overruns compared with the estimate by the PSC staff. Between 1982 and 1984, actual expenditures on Nine Mile Point Two were 50–60 percent greater than the expenditure pattern projected by the PSC staff for a $4.9 billion completion cost. Yet throughout this period, the utilities maintained that the actual costs of Nine Mile Point Two would be substantially less than the staff's $4.9 billion projection.

Did the utilities fail to appreciate the magnitude of the risk of cost increases? Careful examination of the record of the PSC proceeding suggests that the continuation of massive cost overruns has not been a complete surprise to the utilities, despite their sworn testimony that the plant could be completed for $3.7 billion. The clearest indication that the utilities had little faith in their own cost estimates was their alarmed reaction to the request that they provide a "cap figure." A cap figure was proposed as the amount of total plant costs that each co-tenant in the Nine Mile Point Two project would be satisfied to receive from ratepayers if the utility were forced to absorb all costs above the figure while collecting the difference if actual costs proved to be less. The only cap figure proffered by any utility was a figure of $6 billion suggested by Long Is-

FIGURE 12.7 Actual versus Projected Spending on Nine Mile Point Two, 1981–1984

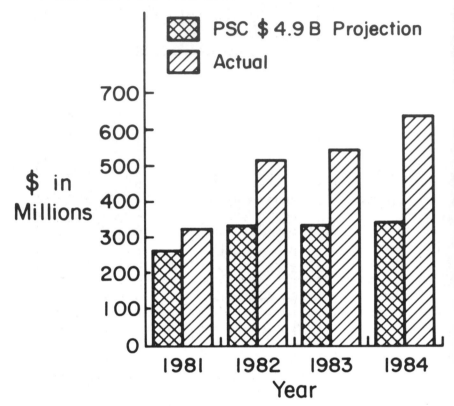

Source: New York Department of Public Service, "The Economic and Financial Implications of Nine Mile Point Nuclear Station Two and Its Alternatives," September 1981, p. A-5; New York Power Pool, Section 5-112 Reports, 1981, 1982, and 1983. Revised 1983 budget from "Monitoring Report No. 9 on Nine Mile Point-2 Nuclear Power Plant," New York Department of Public Service, November 18, 1983.

land Lighting Company's (LILCO) chief financial officer.[8] Given that the utilities' current estimate of completion costs of Nine Mile Point Two is $5.35 billion, the figure proposed by LILCO's executive appears to reflect an accurate appreciation of the potential for cost overruns. Unfortunately, the PSC did not adequately perceive the magnitude of this risk when it decided to permit completion of the plant.[9]

Since the commission's decision to approve completion of Nine Mile Point Two, more than $2 billion has been spent on the project (see Fig. 12.8). Yet reviews in 1983 by both the PSC and the New York State Energy Office (SEO) estimate that the net present value of benefits from

FIGURE 12.8   Estimated Cost of Nine Mile Point Two

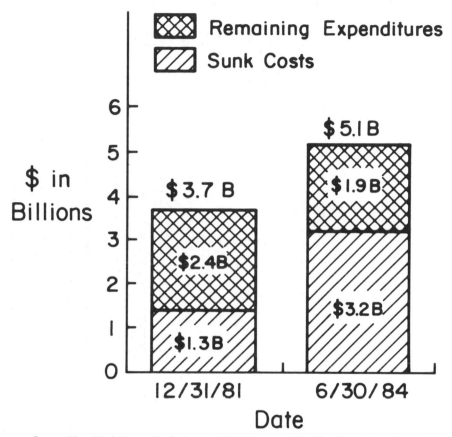

Source: New York Power Pool, Section 5-112 Report, 1982; New York State Energy Office, Nine Mile Point Two Economics Study (1984).

completion of the project—the savings compared with oil-fired generation that was the basis of the "do both" rationale—is approximately $1.5
billion (see Table 12.1).[10] Thus, their own data now establish that the
PSC made a costly error when they approved completion of Nine Mile
Point Two. It is clear that had they stopped the project instead, both the
utilities and their ratepayers would be better off today.

The EDF did an updated study of the economics of Nine Mile Two
and Shoreham in January 1984.[11] Even though the sunk costs of each of
these projects had increased dramatically, the study showed that investments in conservation and renewable alternatives would still be a viable economic alternative to the plants. The study found that if both plants
were canceled and alternatives substituted in their place, ratepayers

TABLE 12.1 Nine Mile Point Two: Two Economic Studies

	PSC Study *April 1984*	*SEO Study* *May 1984*
Estimated net benefits of completion over cancellation	$1.5 billion	$1.6 billion
Cancellation date assumed in study	July 1, 1984	January 1, 1985
Expenditures from 1982 PSC decision to cancellation date	$1.8 billion	$2.3 billion

Source: Memorandum to the New York Public Service Commission, "Staff Economic Analysis of Nine Mile 2," April 26, 1984; New York State Energy Office, Nine Mile Point Two Economics Study (May, 1984).

would realize net benefits of $1.1 billion even after repaying all the sunk costs of the projects.

The SEO reviewed the EDF study and concluded that a selected set of conservation alternatives would be cheaper than completion of the Nine Mile Point Two project.[12] However, the SEO study severely criticized the EDF because it "failed to consider the economics of the conservation and renewable resource alternative as a supplement to Nine Mile 2." The study further maintained, "There is no apparent reason why the co-tenant utilities could not invest in conservation and renewable resources in addition to Nine Mile Point 2 and realize the resultant savings, as well as savings accruing from the Nine Mile Point 2 investment."[13] The SEO reaffirmed its endorsement of completion of Nine Mile Point Two on the ground that the project was still expected to offer some net economic advantage over existing oil-fired generation.

Thus, New York authorities continue to fall into the trap of the "do both" argument, even though experience has shown that it sacrifices pursuit of the most economic alternative in favor of an alternative that continues to grow less economically viable. No rational investor would be content to invest his or her capital in securities that provide lower returns and greater risk than competing investment alternatives simpy because the return is greater than zero. Yet this is precisely what the New York regulatory authorities continue to permit the utilities to do.

THE CONSERVATION DECISION

Although the New York PSC approved completion of Nine Mile Point Two, the commission also launched a special proceeding to require all seven of the state's investor-owned utilities to consider the EDF's conservation proposals. The utilities initially proposed that hearings be postponed for two to three years while they performed studies to assess the impact of providing financial incentives for conservation. They maintained that they knew little about appliance end-uses in their service territories and they continued to predict that customers would not respond to conservation incentives. The administrative law judge ordered them to commence studies immediately and scheduled hearings to consider the cost-effectiveness of conservation investment.

The utilities conducted a statewide survey of appliance end-use patterns and three utilities experimented with pilot programs as described above. The PSC also sponsored a symposium that brought representatives of out-of-state utilities to New York to discuss their experience with utility conservation investment programs.

In April 1983, testimony was filed on the cost-effectiveness of residential conservation measures. Each of the seven utilities used a different methodology and different assumptions concerning the costs and energy savings of the same set of residential conservation measures. Energy savings assumed for certain measures varied by a factor of 6 from one utility to another. One utility estimated that low-flow showerheads would cost an average of $21.50 each, while another utility had actually procured them for its pilot program for $5.08 each. Rochester Gas and Electric projected that it would cost $24.00 to process each rebate (including $7.25 to write each rebate check), although New York State Electric and Gas had incurred administrative costs of only $2.00/rebate in its pilot program.

Despite their wide divergence in assumptions and methodologies, the utility studies generally found that each of the residential conservation measures would produce savings several times greater than its cost. For example, the studies found that each low-flow showerhead would produce savings with a net present value ranging from $214 in Orange and Rockland's service area to $428 in the LILCO's; net savings from each water heater wrap ranged from $11 in the service area of Rochester Gas and Electric to $262 in the LILCO's.

Despite the enormous net resource benefits of conservation investments, all utilities opposed provision of financial incentives to stimulate such investments. Because they refused to credit conservation with significant capacity savings, they claimed that the revenue loss conservation would disadvantage nonparticipants as fixed costs were spread over fewer kilowatt-hours of sales. Curiously, all utilities supported informational programs to promote conservation, even though they had to admit that to the extent that such programs were successful in stimulating conservation they would generate the same revenue loss and have the same adverse impact on nonparticipants.[14]

Although some utilities acknowledged that capacity savings produced by conservation investments could reduce rates to all customers, they generally maintained that conservation could not be given credit for capacity savings unless it could be demonstrated that specific conservation investments would defer planned capacity additions. With both Shoreham and Nine Mile Point Two under construction, the utilities maintained that additional generation capacity savings would be minimal. Although marginal capacity cost estimates routinely are computed for rate design purposes, the utilities argued strenuously that they should be able to use different estimates of marginal capacity cost to compute the avoided costs of conservation.

In November 1983, the administrative law judge presiding over the conservation proceeding released a recommended decision. The decision found that there are no legal barriers to direct utility investments in conservation and that the PSC has ample authority to require New York utilities to pursue conservation investments. The decision states that in light of the continued construction of Shoreham and Nine Mile Point Two, the immediate benefits of conservation investments are substantially reduced. However, the decision concludes that as avoided costs continue to rise, full-scale conservation investment programs will become more economically viable. Thus the decision outlines a staged plan for utilities to develop data, experience, and managerial expertise to implement full-scale conservation investment programs.

The PSC adopted most aspects of the recommended decision. The commission ruled that New York utilities must treat conservation investments on an equal footing with investments in new generating capacity. The PSC directed the utilities to spend 0.25 percent of their revenues to implement conservation programs so as to develop the experience and managerial expertise to pursue subsequent, full-scale conservation investments.

Unfortunately, the utilities have remained opposed to conservation investments, and their initial filings in response to the commission's decision have been very disappointing. Figures 12.9–12.14 summarize the conservation programs proposed by six of the seven utilities in initial compliance filings. Rochester Gas and Electric is not included because its compliance filing did not include expenditure figures. As Fig. 12.15 demonstrates, most of the utilities' planned expenditures for conservation are for information programs rather than for programs involving direct utility investments in conservation. Although some utilities propose to offer rebates to their customers for purchasing conservation devices, none of the utilities is planning to offer a free installation program, despite the demonstration by Niagara Mohawk's pilot program of the dramatic results free installation programs can achieve.

INCENTIVES FOR UTILITY INVESTMENT IN CONSERVATION

The key difficulty in developing successful programs of utility investment in conservation will involve changing utility attitudes toward such investment programs. If a utility remains opposed to direct investments in conservation, it will be very difficult to get the utility to operate a successful conservation program. Unfortunately, most of the New York utilities have shown few signs of altering their opposition to such progams.

In order to remove some of the disincentives to utility investment in conservation, the EDF proposed a balancing account mechanism to pre-

FIGURE 12.9   Central Hudson Conservation Program

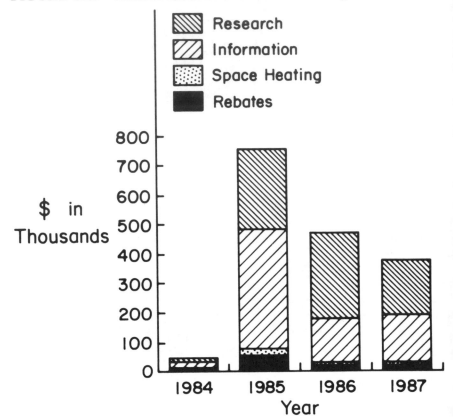

Source: 1984 Compliance Filing of Central Hudson Gas & Electric in New York Public Service Commission Case No. 28223, Proceeding to Inquire into the Benefits to Ratepayers and Utilities from Implementation of Conservation Programs that Will Reduce Electric Use (August, 1984).

vent short-run utility revenue losses from arising because of additional utility financed conservation. Although the administrative law judge adopted the EDF's proposal, the PSC in its final decision rejected it.

Another significant disincentive to utility investment in conservation is provided by the current structure of the federal tax code. The tax system offers substantial subsidies to utilities for construction of central station power plants, which are not generally available for investments in conservation. The investment tax credit, the accelerated cost recovery

FIGURE 12.10   Consolidated Edison Conservation Program

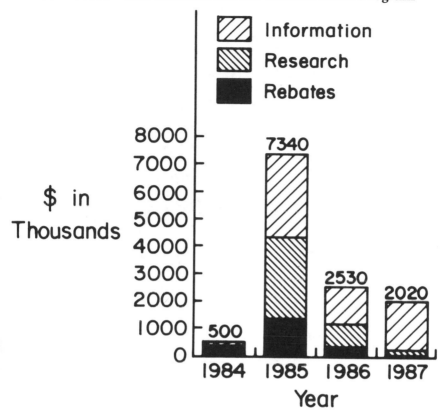

Source: 1984 Compliance Filing of Consolidated Edison Company in New York Public Service Commission Case No. 28223 (August, 1984).

system, and the use of tax exempt pollution control bonds all permit utilities to avoid or postpone billions of dollars in federal taxes for power plant construction projects. These subsidies for power plant construction are estimated to cost the federal treasury $12 billion annually.[15]

The U.S. Department of Treasury's original tax reform proposal[16] would eliminate many of the distortions caused by current tax incentives for power plant construction. If adopted, these reforms (including the elimination of federal tax credits for conservation and renewable energy investments) will go a long way toward establishing a "level playing field" for utility investment decisions.

FIGURE 12.11 LILCO Conservation Program

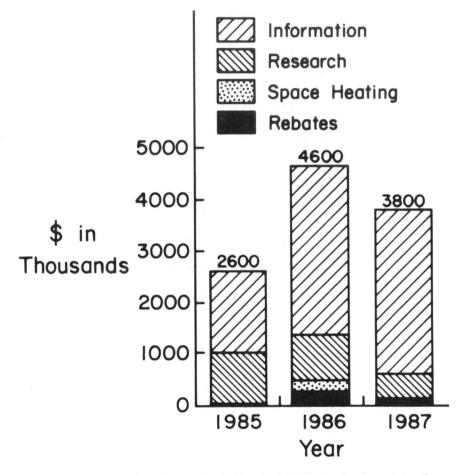

Source: LILCO 1984 Compliance Filing in New York Public Service Commission Case No. 28223 (October, 1984).

FIGURE 12.12   Niagara Mohawk Conservation Program

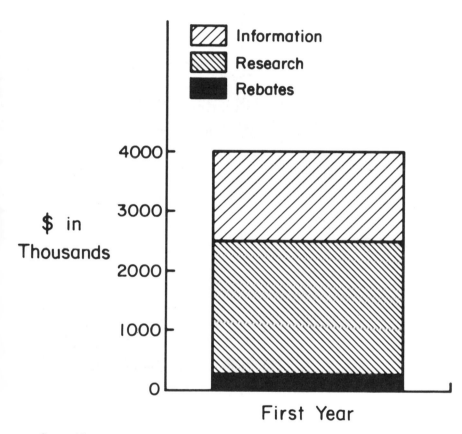

Source: Niagara Mohawk 1984 Compliance Filing in New York Public Service Commission Case No. 28223 (August, 1984).

FIGURE 12.13   NYSEG Conservation Program

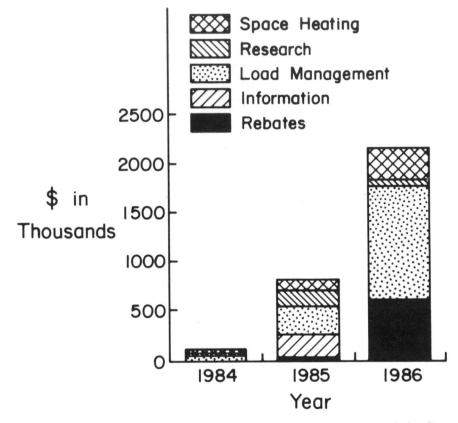

Source: 1984 NYSEG Compliance Filing in New York Public Service Commission Case No. 28223 (August, 1984).

FIGURE 12.14   Orange & Rockland Conservation Program

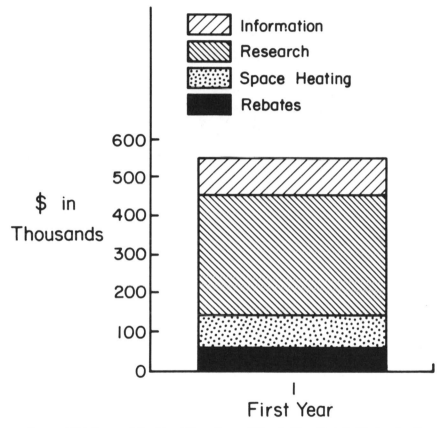

Source: 1984 Orange & Rockland Compliance Filing in New York Public Service Commission Case No. 28223 (August, 1984).

FIGURE 12.15 Size of Conservation Programs and Amount of Direct Investments

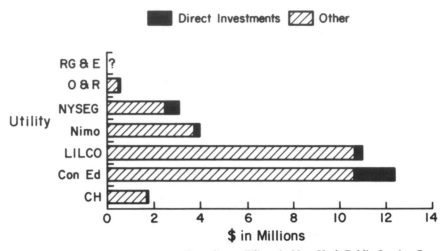

Source: New York Utilities 1984 Compliance Filings in New York Public Service Commission Case No. 28223.

CONCLUSION

New York state today provides an excellent illustration of the consequences of utility investment policies that failed to consider alternatives to construction of central station power plants. New York's utilities are saddled with the two most expensive nuclear power plant construction projects in the nation after regulatory authorities rejected proposals for conservation and renewable energy alternatives to replace these plants.[17]

Although there is some reason for hoping that New York's regulatory authorities have learned a lesson, the utilities' initial response to the PSC's conservation decision indicates that their attitudes may be slow to change.[18] California's utilities did not embrace conservation investment programs enthusiastically until after state regulators had penalized them financially for failing to do so. New York regulators may ultimately have to adopt a similar approach.

NOTES

1. The story of EDF's efforts to persuade the California utilities to embrace direct investments in conservation and renewable energy sources as alternatives to central station

power plants is described in detail in David Roe, *Dynamos and Virgins* (New York: Random House, 1984).

2. The EDF alternative plan and the economic analyses supporting it are outlined in detail in Environmental Defense Fund, *A New Alternative to Completing Nine Mile Point Unit 2 Nuclear Station: Economic and Technical Analysis* (1981).

3. New York State PSC Case 28223, Proceeding to Inquire into the Benefits to Ratepayers and Utilities from Implementation of Conservation Programs That Will Reduce Electric Use.

4. Lawyers for Rochester Gas and Electric even argued during the conservation proceeding that the company's certificate of incorporation would be unconstitutionally impaired if it were directed to invest in conservation because its corporate charter only empowers it to sell electricity (ibid.).

5. Thus, in the conservation proceeding, the utilities found themselves in an uncomfortable position. They opposed utility investments in conservation on the grounds that any conservation in their systems would harm nonparticipating ratepayers while vigorously defending their commitment to informational programs to promote this very "evil."

6. Although Niagara Mohawk maintained that their pilot programs lasted for several months, customers actually had less than four weeks to respond to the utility's rebate or free installation offers.

7. For example, during the conservation hearings attorneys for Rochester Gas and Electric spent considerable time presenting a parade of unusual lamps and lighting fixtures that could not use flourescent bulbs. When questioned, however, they were unable to establish that the fixtures that they displayed were in widespread use. In fact, the Statewide Residential Appliance Inventory, a survey undertaken specifically for the conservation proceeding, found that virtually all households in New York could convert at least one incandescent bulb to fluorescent and that the majority of households could make two or more conversions.

8. Although the commission made a formal request that each utility provide it with a "cap figure," the utilities refused to comply.

9. While the PSC did adopt an incentive rate of return plan that penalized the utilities for cost overruns above a $4.6 billion completion cost, it is doubtful that completion of Nine Mile Point Two would have been approved had the commission accurately appreciated the risk of continued cost overruns.

10. New York State Energy Office, *Nine Mile Point Two Economics Study* (May 1984).

11. Environmental Defense Fund, *The Positive Alternative to Completing Shoreham and Nine Mile Two* (1984).

12. New York State Energy Office, *Nine Mile Point 2 Economics Study, Phase III Report* (September 1984).

13. Ibid., p. 50.

14. Each of the utilities admitted that it had not done any study of the impact of their informational programs on conservation decisions by their customers. When directed by the administrative law judge to determine the potential impact of informational programs on nonparticipants, the utilities were unable to do so.

15. Richard Morgan, "Federal Energy Tax Policy and the Environment" (Washington D.C.: Environmental Action Foundation, April 1985).

16. *Tax Reform for Fairness, Simplicity and Economic Growth: The Treasury Department Report to the President* (Washington, D.C.: U.S. Treasury, 1985).

17. Although the EDF's presentation in the Nine Mile Point Two proceeding represented the most comprehensive case ever made in New York for the economics of conservation and renewable alternatives, intervenors in a LILCO rate case in 1980 proposed a conservation alternative to Shoreham. They projected that a program of investments in residential, commercial, and industrial conservation could displace more oil than Shoreham at a

cost substantially less than the $1–$1.5 billion needed to complete Shoreham. LILCO rejected this alternative, arguing that it was inappropriate and unreliable and that only $500 million would be needed to complete Shoreham at a total cost of $2.2 billion. The current cost estimate for completion of Shoreham exceeds $4.1. billion.

18. For example, Consolidated Edison (ConEd) has recently asked the PSC to relieve it of its obligation to spend $11 million on pilot conservation programs in 1985 on the ground that it cannot prudently devote that amount to conservation spending. Although the PSC rejected ConEd's request, ConEd continues to view the conservation decision as a one-shot nuisance that will not have a major, long-term effect on corporate investment decisions.

13

An Energy Traditionalist Talks to the Energy Nontraditionalists about Electricity and Nuclear Energy——
Alvin M. Weinberg

When nuclear energy was born some forty-five years ago, I was a young physicist filled with brilliant visions of an electrical world based on fission of uranium. I was an unabashed technological optimist who saw nuclear fission as the instrument for achieving H. G. Wells's *World Set Free*. At that time advocates of windmills, dams, and other small electricity sources were the energy traditionalists. I certainly would have been regarded then as an energy nontraditionalist.

How things have changed in forty years! We "nukes" are now considered hopelessly old-fashioned. The new energy gurus by and large reject electricity (except where it is absolutely necessary); they like decentralized energy systems and they abhor nuclear electricity. For them, whatever electricity we shall need (which is not very much) can be supplied by a combination of conservation, small-scale decentralized devices, and co-generators: the day of the 1,000-megawatt (MW) electrical plant, whether coal *or* nuclear, has passed.

I shall explain why, despite the great influence the nontraditionalists have had on energy policy, I believe the traditionalists' view will ultimately prevail.

THE TRADITIONALISTS' VIEW OF ENERGY, PARTICULARLY ELECTRICITY

The traditionalists' view of energy can be summarized as follows:

Proposition 1. The society will become more, not less, electrified.
Proposition 2. An electrical society is economically efficient, energy efficient, and environmentally benign.

Proposition 3.   Electricity will continue to be generated predominantly in large, central plants, many of which will be nuclear.

Proposition 1: Electricity demand will continue to increase. Electricity continues to penetrate, both in the United States and in the world, even in the face of an overall trend toward conservation. If conservation is measured by the decrease in the ratio of *energy* to gross national product (GNP), then the world, and particularly the U.S., has performed surprisingly well. In the decade from 1970 to 1980, the ratio of energy to gross world product fell by more than 10 percent; in the U.S. this ratio fell by almost 20 percent. At the same time, the ratio of *electricity* to GNP has *increased* by about 13 percent during this decade, both for the world and for the U.S. (see Fig. 13.1). Thus the fraction of primary energy that is converted to electricity increased by an astonishing 30 percent during the decade for both the entire world and for the U.S. (Fig. 13.2), and now stands at about 35 percent in the U.S. The total amount of electricity used in the U.S. has grown by an average of 7 percent a year, excepting the years of the oil crisis, when its growth fell to less than 2 percent. In 1983 it grew by almost 8 percent. All but the most revolutionary nontraditionalists agree it will continue to rise for the next few decades—perhaps by 3.0–3.5 percent a year,[1] perhaps by only 2.5 percent a year.[2]

Why has the marketplace thus far ruled against the nontraditionalists' view that electricity would gradually be displaced by decentralized nonelectrical modes? Two reasons stand out: first, although the average price of a kilowatt-hour (kWh) of electricity in the U.S. rose from 1.9¢ in 1973 to 5.8¢ in 1982, the ratio of the price of electricity to the price of other energy vectors fell in this period. Indeed, in the U.S. the ratio of the price of electricity to the price of oil (per end-use BTUs) is, on average, below 3:1. At this price ratio, electric-resistance space-heating, because of its greater end-use efficiency, becomes competitive with the direct use of oil! Of course, for New York City, which is burdened with extremely high electricity rates, the ratio is much less favorable.

A major reason for the relative stability of electricity prices is that electricity is priced at its average, not its marginal price. Much of our electricity comes from plants that were bought at low cost or that have been fully amortized. This stabilizing effect is accentuated by a gradual shift now occurring in electrical generation away from oil and toward coal and uranium (both of which have much lower fuel costs (per BTU than oil). Thus the longevity of central electrical generating systems is an important determinant of the future cost of electricity.

Throughout the utility industry one finds great interest in extending the lifetime of old plants: because new plants are so expensive, economics

**FIGURE 13.1   Time Trends of Primary Energy and
Electricity Use in Relation to GNP**

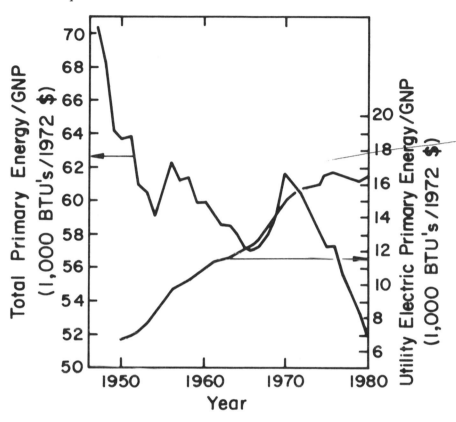

Source: A.M. Weinberg & C.C. Burwell, "The Rediscovery of Electricity," *Proceedings of the American Power Conference,* 1982, v. 44, p. 13.

drives utilities to extend the life of their older plants, both nuclear and coal. The Calder Hall reactors, which have exceeded their original 25 year licensed operating lives, have been relicensed for another 5 years. At the Institute for Energy Analysis (IEA), we have concluded that most of the current nuclear power plants will be operable for some time after their originally expected lifetime expires. Once a nuclear plant has been amortized, the price of electricity decreases to around 1.5¢/kWh in current dollars for fuel plus operating and maintenance costs. That these considerations are not merely theoretical is reinforced by history: A Newcomen engine built in 1791 operated in England until 1918!

The second reason for electricity's penetration has to do with its efficiency as a factor of productivity. Many process industries in the United

FIGURE 13.2   Fraction of Primary Energy Used for Electricity over Time

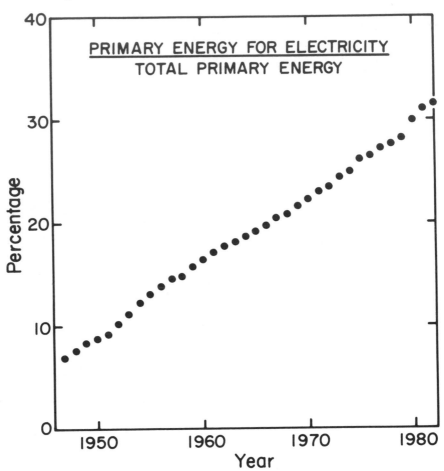

Source: R. Rotty, "Electrification: A Prescription for the Ills of Atmospheric CO_2," *Nuclear Science and Engineering*, 1985, v. 90, p. 467.

States are electrifying because they thereby lower their production costs. As an example of this trend, more than 30 percent of all steel made in the U.S. now comes from electric furnaces, compared with 10 percent some twenty years ago. If the economy continues to grow, I would anticipate an increase in electrical demand, though I am unprepared to specify by what amount. In any event, even if total energy grows very slowly, I would argue that the demand for electricity will continue to grow substantially.

Proposition 2: An electrical society is more economically efficient, energy efficient, and environmentally benign than a nonelectrical society. I have already pointed out that process industries are electrifying because this makes economic sense. That electric heat may be energy efficient also flies in the face of most conventional wisdom: everyone knows that a power plant converts only 35 percent of its primary energy into electricity. But at the point of end-use, the situation is reversed. Because it can be controlled so well, electricity delivers its heat with 100 percent efficiency and only to the precise point where it is needed. A study at the IEA of the relative efficiency of home heating by oil and by electricity showed that at point of end-use, the energy (in BTUs) required to heat a house electrically was less than 40 percent of that required if the heat was supplied by oil. This efficiency at point of end-use almost made up for the thermodynamic inefficiency at the point of generation.

In process industries, the efficiency advantage is even clearer. For example, to make a ton of steel electrically (from scrap) requires 75 percent less primary energy than that required to make it by conventional means. And the electric steelmill is nonpolluting at the point of end-use! We believe such improvements explain at least part of the remarkable increase in use of electricity at the same time total energy usage is falling—in short, that electricity per se is a conserver of energy. Many nontraditionalists have failed to recognize this fact. They foresee reduced electricity demand because of conservation measures: improved home insulation, more efficient lighting and motors. What they fail to consider arc the entirely new uses of electricity, particularly in the process industries, where its universal applicability, its exquisite controllability, and its cleanliness make electricity the energy form chosen; or its possible new uses, still speculative, such as electrification of agriculture, or electric vehicles.

Electricity is in most ways environmentally benign: this follows directly from its energy efficiency at point of end-use. Certainly at point of end-use, pressure to electrify comes from producers plagued with messy fossil fuel-fired process heat systems: electrically melted glass creates no pollution, but gas-melted glass creates dirty scrap that must be cleaned up. One large scrubber to remove sulfur oxides on a large fossil fuel-fired power plant is more economical than are small scrubbers on many small industrial boilers. And, in any case, properly operating nuclear plants are our cleanest source of all for thermally-generated electricity.

Proposition 3: Electricity will continue to be generated predominantly by large, central plants, many of which will be nuclear. If my argument about longevity of existing plants is correct, then, a fortiori, proposition 3 is valid. However, I would insist that the current fashion and popularity of so-called modular plants may be a passing reflection of the financial plights of many utilities—that the scaling laws are not so easily repealed. One piece of evidence of this is the trend in co-generation by industrial users. As seen in Fig. 13.3, industrial co-generation has fallen by a factor of almost 2 between 1969 and 1982. This must reflect the economic reality: that electricity generated centrally is cheaper than electricity generated as a by-product in small industrial installations.

At the moment this trend may be reversing. For example, in the Houston area, the Dow Chemical Company and other companies are installing huge co-generating plants—one produces 600,000 kW. The excess electricity must be taken by Houston Power and Light, according to the

FIGURE 13.3 Time Trends of Industrially Generated Electricity

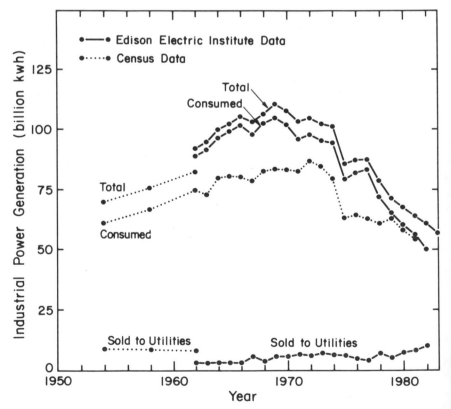

Source: C.C. Burwell, "Electrification Trends in the Pulp and Paper Industries," Report No. ORAU/IEA 84-11M 1984 (Oak Ridge Associated Universities).

Public Utility Regulatory Policy Act (PURPA), at the full avoided cost. This situation has imposed a tax on other users, both residential and industrial, despite its manifest unfairness. To what extent co-generation would be competitive without the PURPA subsidy remains to be seen.

Whether or not we will accept nuclear power again is not so clear. Improvements, both technical and institutional, will be needed for this to happen—all of them aimed at increasing the public's acceptance of nuclear power and diminishing the cost of nuclear plants. About half of all Americans now oppose nuclear energy because they believe the waste problem is intractable; they are afraid of reactor accidents; they worry about proliferation; and they associate nuclear plants with higher electricity costs. Utility executives in this country have turned away from nuclear plants because they are too expensive, their costs are indeterminate, and their regulation is cumbersome and uncertain. I shall deal with each of these points briefly.

Waste disposal. Despite the public's apprehension, progress in this area is being made. Most important, the Nuclear Waste Policy Act of 1982 is now in place. One mill (0.1¢) is assessed against every kWh generated in a U.S. nuclear plant, the money to be used to carry out the provisions of the act. Several possible waste disposal sites are now being assessed. In France, West Germany, and Sweden, tangible progress is being made: reprocessed wastes in France are being incorporated into borosilicate glass; Sweden has almost finished its retrievable spent-fuel storage facility; West Germany is operating its salt mine repository. In the meantime, technical improvements are being made—for example, Oak Ridge National Laboratory (ORNL) has developed a lead-iron phosphate glass that is a thousand times more resistant to groundwater attack than is the conventional borosilicate glass. The package, even in groundwater, would lose no more than 10 microns of surface a year: in 100,000 years the penetration would be at most 1 centimeter, and probably much less. I estimate that during the next decade, enough evidence of successful waste disposal strategies will have been amassed to persuade people that wastes *can* be and *are* being disposed of safely.

Reactor safety. Sparked by the accident at Three Mile Island Two (TMI-2) both technical and institutional improvements are being made. Backfits mandated by the Nuclear Regulatory Commission (NRC) after TMI-2 have now been incorporated in existing reactors; and a recent study of "accident precursors" performed by ORNL and Science Applications, Inc. has indicated that during 1980, the probability of a core meltdown was about 15×10^{-5}/reactor year. This agrees fairly well with Rasmussen's predicted core meltdown probability of 5×10^{-5}/reactor year, with an uncertainty of a factor at least 5. If we take 15×10^{-5}/reactor year for the average core meltdown probability, this being a semi-empirical estimate, we can say that, with 120 reactors operating in the U.S. during the next fifteen years, the expected probability of another core meltdown would be around 0.3. The likelihood of anyone being hurt in such an accident is much lower: one must remember that at TMI-2 less than one-millionth of the radioactive iodine contained within the reactor actually escaped to the environment.

I believe this expected probability is an overestimate, partly because as weaknesses in the hardware and the operating cadre are identified, they are being remedied; and because the institutions involved in nuclear energy are being strengthened. Perhaps most notable is the utility-funded Institute of Nuclear Power Operations (INPO). The INPO establishes and enforces standards of excellence in the operation and construction of nuclear power plants. In this respect it reinforces the actions of the NRC. The INPO also serves as a clearinghouse for the exchange of information related to safety. Weak points in design and operation gradually get fixed: indeed, there is little doubt that reactors today are safer than were reactors at the time of TMI-2.

Whether they are safe enough is hard to answer. In any case, reactors are being developed that are significantly safer than the existing ones. The development proceeds along two paths: improvements in reactors where the safety depends on active systems, and development of passively or inherently safe reactors. Today's reactors are actively safe—that is, they rely on active intervention of electromechanical devices to prevent damage if cooling is lost. By adding redundant systems, one reduces the probability of accident—for example, Sizewell-B, a pressurized water power reactor proposed for England, has a core meltdown probability of around 10^{-6}/reactor year; the advanced pressurized water reactor being designed by Westinghouse in collaboration with Mitsubishi, and the advanced boiling water reactor of General Electric with Toshiba and Hitachi both have core meltdown probabilities around 10^{-6}/reactor year also.

Inherently safe reactors are also being discussed seriously, and the U.S. Department of Energy (DOE) has recognized and is supporting, albeit on a small scale, their development. The two most prominent, the Swedish Process Inherent Ultimate Safety (PIUS) and the modular high temperature gas reactor, are essentially immune to core damage. The PIUS reactor achieves this with its huge concrete pressure vessel filled with borated water; its pressurized water reactor is immersed in the pool, and a clever hydraulic lock allows the pool water which contains a chain reaction quencher, boron, to rush into the reactor if the cooling is disturbed in any way (Fig. 13.4). The modular high temperature gas-cooled reactor operates at such low power density that it can cool itself by radiation even if all forced cooling is lost. These two ideas introduce a new phase of reactor development: nuclear power where safety is deterministic rather than probabilistic, and where chances of an accident occurring are as likely as those of a large asteroid striking the earth!

FIGURE 13.4 Schematic Cross-section of the PIUS Reactor

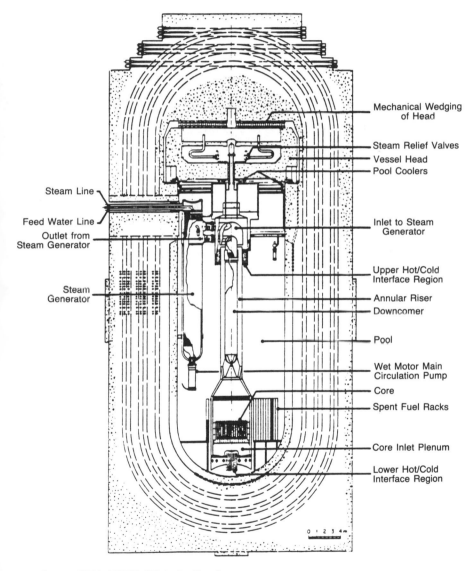

Source: ASEA-ATOM, Västerås, Sweden.

Proliferation. Though the public is generally unconcerned about the proliferation issue, many of the articulate opponents of nuclear power regard proliferation as the major objection to nuclear power. Yet the record so far belies this assessment: none of the six nations known to possess nuclear bombs obtained the plutonium for its first bomb from power reactors. Forty years ago, most people would have predicted dozens of countries with nuclear weapons by now, yet the actual number is far less—there are six certainly, three or four others may be working on bombs. Thus, contrary to the conventional view, I would say the nonproliferation regime has worked better than had been expected.

We cannot prevent proliferation. We can further reduce the connection between bombs and power reactors—in short, strengthen the nonproliferation regime represented by the Non-Proliferation Treaty, the International Atomic Energy Agency, the London Suppliers Club, and the Treaty of Tlatelolco (Mexico). One key idea now being discussed is adoption of the Soviet used-fuel policy: all fuel in Soviet-built reactors must be returned to the Soviet Union. What the Soviets do with this spent fuel is unclear, but any plutonium in these fuel elements can hardly contribute to horizontal proliferation. The Soviet scheme not only eliminates proliferation, it also rids the participating country of its high-level wastes. Thus, in a recently published study, the Washington Institute for Values in Public Policy and the Institute for Energy Analysis have proposed that the U.S. offer to take care of high-level wastes as an incentive for countries using U.S. reactors to return their unprocessed spent fuel.[3]

Economics. Many nuclear power projects have been economic disasters: unless utilities can be convinced that nuclear plants can be built on schedule and to cost, nuclear plants will not be built in the U.S. Yet, there is no law of nature that prohibits nuclear plants from being built on time and to cost: in France, Taiwan, and Japan, exactly this is happening. Whether it can also happen in the U.S. I cannot say; no utility is currently planning to build a new nuclear plant. But over the next fifteen years, we may well learn what the French, Taiwanese, and Japanese have learned. Only a portion of any improvement in meeting schedules will come from licensing reform: we will need standardized, safer plants, which are recognized as such by the NRC; we will need more architect-engineers who can build nuclear plants as quickly as St. Lucie (68 months) or River Bend (62 months); or perhaps more broadly, we may need to consolidate our nuclear enterprise—perhaps, as the Atomic Industrial Forum suggests, form consortia that both build and operate the plants. But above all, we must operate the existing plants safely and reliably so that the public regains its confidence in nuclear power. With regained confidence, cheaper nuclear plants are possible; without the public's confidence, no nuclear plants are possible.

THE NONTRADITIONALISTS' VIEW OF ENERGY, ESPECIALLY ELECTRICITY

The nontraditionalists, who range from Amory Lovins to some Wall Street utility financial analysts, view the remarkable reduction in total energy demand (from a high of 81 quads in 1979 to 71 quads in 1983) simply as evidence that conservation works: our cars are more efficient, our houses are tighter, industry has insulated its process lines. They attribute the penetration of electricity to a distorted price structure: were electricity priced at its marginal cost, its price relative to other fuels would currently be much higher, and our use of electricity would not be as large. Where the traditionalist sees the penetration of electricity as being partly responsible for the lowering of overall energy demand, the nontraditionalist sees this penetration as an incidental by-product of the aforementioned distortion in pricing. Were electricity properly priced, its use, too, would have fallen, and we would currently be saving even more primary energy.

The nontraditionalist therefore views the energy future as being dominated by conservation: the residual uses of electricity, which the more extreme nontraditionalists believe to be very small, will then be handled by the dispersed, small-scale systems, notably fuel cells, photovoltaics, and co-generators.

Conservation. There is little doubt that conservation measures, which have been largely price-induced, have been a great success. The energy to GNP ratio has fallen by 20 percent since the oil crisis. Is this decrease an entirely new trend, sparked by the oil crisis, or is it a continuation of a trend that can be traced back to the 1920s? As judged by the energy to GNP ratio, what we see is not all that unusual—the economy has been becoming more energy efficient since 1920 (Fig. 13.1). At least some of the lower overall energy demand reflects a lower GNP, rather than an unusually large increase in the energy efficiency of our economy.

Most of the conservation measures that nontraditionalists focus on seem to be in residential heating: many houses have been tightened. The flip side of this trend is the increase in internal pollution—and the possible increase in lung cancer. Henry Hurwitz of General Electric has estimated that if all houses met the specifications for insulation recommended by the DOE, then the increased level of indoor radon would, according to the linear hypothesis, induce 20,000 additional cases of cancer a year.[4] Air-to-air heat exchangers eliminate most indoor pollution— yet homeowners, especially those with low incomes and poorly maintained houses, have little financial incentive to install such devices.

Among the more efficient electricity-saving devices, the Philips holmium-activated fluorescent bulb deserves mention. Each bulb costs

$22 in the U.S. (less in Europe), lasts 7,500 hours, and uses only 18 watts to match the light output of a 75-watt ordinary incandescent light. Thus the 428 kWh *saved* during the 7,500 hours of operation cost about 5¢/kWh: if electricity costs more than 5¢/kWh, the lamp pays for itself; if electricity costs less than 5¢/kWh, it does not. The average cost of electricity in the U.S. in 1983 was 5.8¢/kWh. One Philips lamp saves 57 watts; thus to save 1 kW by the use of such lamps costs $380/kW; on the other hand, this expenditure must be made every year (for a lamp burning at 90 percent load factor).

The main issue for the future is how much more conservation is likely to take place; for electricity, we must ask whether improved efficiency in meeting existing demands for electricity (by air conditioners, etc.) will overcome the unmistakable shift among process industries toward electricity.

Photovoltaics. I have found it very hard to ascertain exactly how much a kilowatt-hour from photovoltaics now costs. At Robert Redford's Institute for Resource Management (IRM) conference on the electric utilities in October 1984, J. Caldwell of ARCO, the largest commercial producers of photovoltaics, reported that a kilowatt-hour of photovoltaic electricity now sells for 50¢—presumably with the tax credit included. Others who manufacture photovoltaics put the figure at perhaps twice this. Clearly, photovoltaics have a *long* way to go before they can match the cost of electricity from a $2,500/kW nuclear plant. Moreover, the photovoltaic does not produce firm power: unless practical storage systems are developed, one cannot count on photovoltaics to support a large share of the electricity market as reliably as central station units.

Co-generation. Co-generation has been the darling of the nontraditionalists because of its greatly improved thermal efficiency: instead of 65 percent of the heat in the primary fuel being wasted, most of that heat is used. But this thermal saving is not without costs:

- Co-generating plants are small: they sacrifice economy of scale for overall efficiency.
- Co-generating plants require a firm market for the electricity they generate, if heat is their primary product; or a firm market for heat, if electricity is their primary product.
- Co-generators usually use gas or oil, not coal or uranium. If our energy policy still aims at reducing our use of oil, co-generation with oil defeats this aim.
- Finally, though co-generators use all the heat, theoretically it is by no means clear that electricity should not have supplied those heating needs directly.

Until 1983, the amount of co-generated electricity produced fell steadily, largely because small power plants were, per unit output, too expensive. Sparked by PURPA, co-generation has spurted ahead, but at the cost of a distorted pricing system for electricity, with individuals paying the PURPA-mandated subsidy for the electricity that co-generators sell to their utilities.

Fuel Cells. Fuel cells, which convert natural or synthetic gas directly into electricity at an astonishing efficiency of around 45 percent, are now operating in a utility grid in the city of Tokyo. An earlier 11-MW version of these fuel cells has been installed in New York City, but when Consolidated Edison (ConEd) tried to generate power from the stack in April 1984, it was found that the phosphoric acid electrolyte had migrated from the electrodes, thus preventing the cells from operating. ConEd and United Technology Corporation have asked Congress for funds to retrofit the New York City unit with improved stacks. Between the Tokyo and the New York City experience, one can conclude that fuel cells can provide blocks of power in the range of 10 MW: in Japan, 50-MW installations are being proposed. Costs at present are \$3,000/kW, which, at 6,000 hours in a year and 20 percent fixed charge, comes to 10¢/kWh for capital; if natural gas at \$5.00/10^6 BTUs is the fuel, the total cost of electricity from fuel cells would be in the range of 15¢/kWh. One hopes that, with a growing market, the cost of fuel cells might fall, perhaps to \$1,000/kW. The big unknown, of course, is the longevity of fuel cells: will they last 2, 10, or 50 years? I regard the answer to this question as a most essential determinant of the eventual place of fuel cells in utility generation of power.

HOW THE UTILITY PRESIDENTS REACT TO TRADITIONALISTS' AND NONTRADITIONALISTS' VIEWS OF THE ENERGY SITUATION

Curiously, utility presidents, public service commissioners, and Wall Street analysts are much taken by the nontraditionalists' arguments. The bitter experience of costly, even canceled nuclear plants has been elevated to a supreme doctrine, the dictum: Don't build large central station power plants. Depend on conservation, on power purchases, on co-generation, and small modular power plants (if one must build)—but build only as a last resort. Wall Street loves utilities that are *not* expanding, especially those that are not building nuclear plants.

Thus utility presidents at present seem to have more in common with environmentalists than with sellers of central station electric plants. As Amory Lovins argued at the IRM conference, increased efficiency in end-use appliances will soften electric demand; any residual electrical additions can be provided by co-generation; large central electric generating plants are obsolete. Several utility presidents at the meeting seemed to agree with this assessment and prognosis. Yet I cannot believe the future requires no more central stations; consider the following:

- Item: Though efficiency (that is, conservation) is improving, electricity use continues to grow. This is the hard, uncomfortable fact that cannot be ignored: *new* uses for electricity are being discovered; I see no reason why this trend should suddenly stop.
- Item: At the IRM conference, several public service commissioners worried that too large a penetration by co-generators could reduce the reliability of electrical service. As I said before, Carnot efficiency is not the only criterion by which a supply option is to be judged—and co-generators without a guaranteed utility buy-back are not always able to match their heating and electrical loads.
- Item: The time frame for utility projections now is very short—five years rather than 20 years. But is it clear that the strategies appropriate for meeting projected demand over the next five years are appropriate for the long term? I am unconvinced of this—the case for expansion of electrical demand seems to me to be as strong as, and perhaps stronger than, the case against expansion of electrical demand.

Can utilities adopt a low-risk strategy that does not place an artificial limit on the amount of electricity we will want in the year 2000, yet that avoids the high cost of small plants? I have already suggested alternatives, some of which are perhaps relevant to New York State. These include integration of new generating facilities, whether coal or nuclear. To me, ten companies each owning one-tenth of a 1,000-MW plant makes more long-term sense than each company building a separate 100-MW plant.

This strategy would require considerable change in the way such generating consortia are put together. One interesting pattern was suggested by J. Geist, president of Public Service of New Mexico, at the IRM conference. A consortium of private companies, not all of them utilities, is proposed to build a system of five 400-MW mine-mouth coal-fired power plants on the Navajo reservation; the power would then be sold to utility companies in the area.

Perhaps the incremental and the large-scale strategies can be combined to the advantage of both. Let utilities expand in two stages: first, by incremental addition either of purchased power or by building small, expensive modules, say fuel cells, or, what are more likely, gas turbines, of 100 MW. When output of these small plants adds up to 500 or 1,000 and a firm demand for the power has been established, replace these expensive, small generators with a single, large, long-lived plant. In this way one minimizes the risk of overbuilding yet preserves the economy of scale of the large plant. I should think this hybrid strategy might emerge eventually—but of course it depends on whether or not the cost of large plants, particularly of nuclear plants, eventually becomes lower than the cost of the modular, generally oil-fired, plants. Given such a scenario, we might see a reenactment of the utility strategy of the 1940s and 1950s: new, highly efficient, less expensive plants would displace old, less efficient plants. But in our example the new plants would presumably be nuclear, or possibly coal; the displaced generators might be small gas turbines or fuel cells or co-generators, all of which wear out relatively quickly and/or use fuel, gas, or oil, whose price, relative to coal and uranium, may well rise in the next 15 years. Of course, this suggestion must be regarded as tentative: each utility has to decide for itself whether increments are better added as small power plants, as purchased power, or as a share of a large plant.

As far as nuclear plants go, none will be built until builders and Wall Street analysts are convinced their cost can be ascertained in advance and that the costs are competitive. Whether these conditions can be achieved with reactors that are improved light-water reactors, or whether we shall need new, inherently safe designs, I cannot say. One hopes that by 1990 this question will have been resolved and that, in the meantime, safer reactors that demand less of regulators and of operators will be developed for commercial use.

Again, in thinking of the long run, one has the impression that what utilities, Wall Street, and nontraditionalists find to be in their best interest—build nothing, or build tiny plants—may not be in the long-term best interest of the society. An important consideration, which I have already mentioned, is the longevity of large plants. If they can be refurbished—say for 10, 20, even 40 years after their design—they will be providing cheap power for our children. Such a time frame will tend to keep the price of electricity low, so low that, even if utilities "overbuild," the price of electricity from these plants will provide a hedge against inflation, especially if the plants are nuclear. This low long-term price of electricity should then lead to a resurgence of electrical demand—a further electrification of the society.

Nontraditionalists cry foul: marginal cost pricing is the economically sound policy. But this argument, I fear, belies a prejudice against big electrical systems. Traditionalists argue: but if replacement is 20–30 years in the future because the plants last so long, then the *discounted* cost of a new plant is not much different and may be lower than the embedded cost of the old plant. So let electricity prices fall, and let the society become more electrified. To me, a nuclear traditionalist, such a society is environmentally benign and economically efficient and preserves the freedom of choice we now enjoy.

NOTES

1. See Department of Energy (DOE), *1983 Annual Energy Outlook* (Washington, D.C.: DOE, May 1984).

2. See Joseph Gustaferro, *U.S. Energy for the Rest of the Country* (Washington, D.C.: DOE, July 1984).

3. Alvin Weinberg, Marcelo Alunso, and Jack N. Barkenbus, eds., *The Nuclear Connection*, (New York: Paragon House, 1985).

4. In a letter to Peter Bradford, the Nuclear Regulatory Commission commissioner, March 27, 1980.

14

A Survey of Alternative Electric Utility Supply and Use Technologies

Frank L. Huband

This chapter surveys innovative electric supply and use technologies, focusing on those that may contribute to New York State's electric power supply over the next few decades. These technologies can be clustered into four groups: (a) "enhanced" use of fossil fuels; (b) renewable energy; (c) nuclear energy; and (d) alternatives to new generation capacity. The penetration into the electric utility marketplace of any new energy technology will depend not only on the characteristics of the technology itself, but in substantial measure on the economic and regulatory implications of that technology. This chapter does not pursue in detail every such implication for each technology but will indicate some factors of special relevance to New York State.

The relative contribution of current energy sources for electric power generation within both New York State and the nation (see Fig. 14.1) is the framework for this discussion. More than half the electricity generated in the United States comes from coal, compared to about one-seventh in New York State. New York State is five times as dependent on oil for electricity production as the nation as a whole; about one-third of the state's electricity generation derives from oil.

None of the sources in Fig. 14.1 have unlimited expansion potential. Coal is dirty; oil is expensive; natural gas is both expensive and has a statutory prohibition on future utility use; nuclear power has problems with both public and financial community acceptance; and hydro-power has a naturally limited expansion potential and negative land use consequences. These problems motivate consideration of alternative energy supply technologies.

FIGURE 14.1   Net Utility Electric Generation by Energy Source

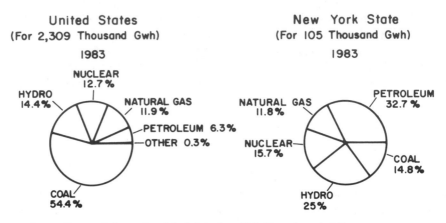

Source: Energy Information Administration, U.S. Department of Energy.

ALTERNATIVE ELECTRICITY SUPPLY AND USE TECHNOLOGIES

Enhanced Use of Fossil Fuels

The use of traditional *fossil fuels*—coal, oil, and natural gas—can be enhanced through a variety of new technologies. Use of these technologies would reduce adverse environmental impacts, improve efficiency of electrical production, or provide new sources of clean fossil fuel. Several of these technologies—co-generation, fuel cells, coal/water slurries, fluidized-bed combustion, coal-bed methane, and coal-gasification combined cycle—appear to have a relatively high near-term potential for commercial development.

Co-generation is not a new technology; in fact, it is not a single technology at all. Co-generation refers to electric power generation in which the primary energy provides not only electricity but also thermal energy for productive purposes. The effect of co-generation is to increase overall energy use efficiency. When co-generation is employed, the electric power is usually generated where the thermal energy is needed. Co-generation was developed prior to World War I. By 1918, about one-half of all electricity produced was generated by the industrial plant that used it,[1] and many facilities employed co-generation. The development of a reliable electric utility industry and electric-generating economies of scale combined to reduce the proportion of electric generation produced by the non-utility industry sector from 47 percent in 1912 to less than 3 percent in 1980. There are factors, however, that have the potential to reverse this trend, including new electric-generating technologies capable of clean and efficient production of electricity in units of small size at competitive cost. Most of the remaining technologies discussed in this section have potential as co-generators. Additionally, the Public Utility Regulatory Policy Act (PURPA) requires utilities to interconnect with and purchase excess electricity from co-generation facilities at the price it would cost the utility to produce that electricity, the so-called avoided cost. In regions where high-cost oil would otherwise be used to produce electricity, the PURPA-mandated avoided-cost price should be attractively high and an incentive to potential industrial co-generators.

The Office of Technology Assessment has projected an ultimate co-generation potential for the U.S. industrial sector alone of 200 gigawatts (GW, or 200,000 megawatts [MW])[2]—representing almost one third of current U.S. electricity generating capacity. Another source estimates that co-generation and small power production will reach 25 GW by 1995.[3] In regions where growth in co-generation capacity seems either socially beneficial or inevitable, regulatory commissions may find it desirable to provide increased economic incentives for public utilities to participate actively in the development of that capacity.

Fuel cells have been characterized by the Electric Power Research Institute (EPRI) as the technology that will be available when the utility industry needs its most.[4] Fuel-cell power plants produce energy by electrochemistry rather than by combustion. Three types of fuel-cell systems are under development: phosphoric acid (PAFC), molten carbonate (MCFC), and solid electrolyte (SEFC).[5] Currently available PAFCs produce electricity at 40–45 percent efficiency from clean fuels such as naphtha or natural gas, in modules of up to 11 MW. These fuel cells produce electricity at temperatures below 200°C. When produced in volume, these cells should cost about $1,000/kilowatt (kW). MCFCs are more advanced than PAFCs, with the potential, when fully developed, to develop 50 percent coal-to-busbar efficiency using a medium-size BTU coal gasifier, or 60 percent efficiency using clean fuels. Because the operating temperatures of MCFCs is high, about 650°C, the by-product heat produced in MCFCs should be usable as industrial process heat. SEFCs are still in the laboratory stage of development. They are projected to operate at temperatures greater than 1,000°C at high efficiencies.

In 1977, the Department of Energy (DOE) and EPRI funded a Consolidated Edison demonstration phosphoric acid fuel-cell plant that was built by United Technologies Corporation (UTC). Because of installation delays and inadequate protection during storage, the fuel-cell stacks deteriorated, and the facility will not be completed. UTC improved the stack design of these cells and produced a 4.5-MW fuel cell for Tokyo Electric Power Company. The plant has operated without problems since starting up in April 1983.[6] The modularity, reduced environmental impact, and estimated 2-3 year time from order to completion should make fuel cells an attractive technology. EPRI projects a 20-35 GW electric utility market by the year 2005.[7]

Coal-water slurry (CWS) technology provides the potential for retrofitting current oil- or gas-fired generators to burn a coal-based fuel. CWS fuels consist of water containing finely ground and cleaned coal from which about one-half the pyritic sulfur has been removed. Estimates for the cost of retrofitting range from $150 to $250/kW,[8] and the cost of CWS fuel is estimated at from $2.80 to $3.50/million BTUs. This compares with about $4.00/million BTUs for gas and $4.50/million BTUs for oil in New York State in 1983. For some types of boilers, however, the CWS retrofit may reduce boiler capacity. An unresolved regulatory issue that could seriously affect the potential for retrofit CWS is whether retrofit of an oil-fired boiler to coal-water slurry capability will be construed by regulators as a major modification, subjecting it to new source performance environmental standards. If so, scrubbers could be required on coal slurry conversions, adding from $250 to $500/kW to the cost of the conversion.

Fluidized bed combustion (FBC) is an improvement in the design of conventional coal-fired boilers that eliminates the need for pollution control equipment. Limestone is injected into the boiler to chemically absorb sulfates and nitrates, preventing these effluents from being released into the atmosphere. In addition, FBC boilers are efficient heat exchangers and operate at relatively low temperatures (815–870°C). At this temperature, noncombustible ash-forming minerals in coal do not melt, thus eliminating the formation of the molten slag that corrodes boiler pipes. FBC boilers can burn a wide variety of fuels, including garbage and all grades of coal. FBC boilers are economical in sizes as small as 100 MW. Atmospheric FBC, in which the combustor operates at atmospheric pressure, has been successfully demonstrated at the pilot plant stage; it needs to be demonstrated on larger-scale plants and utilities. Pressurized FBC, in which the combustor is pressurized at 6–16 times atmospheric pressure, is in an earlier stage of development.

Coal-gasification combined cycle is another technology that can serve as a potentially clean source of electricity. The first demonstration plant for this technology, the Cool Water Facility in California, generating 120 MW of electricity from the combination of a gas turbine and steam turbine, began operation in July 1984.[9] Efficient methods of cleaning the hot intermediate gas without cooling, currently under development, offer potential efficiencies of up to 50 percent for combined-cycle generation, with low sulfate, nitrate, and particulate emissions.

Coal-bed methane (CBM), natural gas that lies within coal seams, is a fossil fuel source that could have great potential nationally and in the Northeast particularly. For the northern Appalachian coal basin, which includes western Pennsylvania, northern West Virginia, and eastern Ohio, this resource has been estimated to be as much as 60 trillion cubic feet (60 quads) of pipeline-grade methane.[10] Depending on ownership, price, Fuel Use Act waivers, and other conditions, CBM could be a relatively inexpensive, abundant energy source for New York State utilities.

Other technologies in the fossil group, such as *coal-fired turbines and diesels, in-situ coal gasification, synthetic liquids from coal,* and *magnetohydrodynamics* (MHD) appear to have substantially lower near-term potential both for New York State and the nation.

Renewable Energy Technologies

Most *renewable energy technologies* rely directly or indirectly on the sun and thus suffer from the diffuse nature of that energy resource. As a result, use of these technologies typically requires large land areas, on the order of 10-20 acres per peak MW.[11] Thus, all other factors being equal, most renewable energy technologies are likely to become economical first in areas where land is abundant and inexpensive. Additionally, because of the higher solar intensity in the Southwest, a solar facility of a given size (other than a wind facility) located in that part of the country would produce 50 percent more energy than if located in New York State. The Southwest's combination of lower land costs and higher solar intensity thus makes that region the most likely demonstrator of utility-scale solar technology.

Solar photovoltaics represent the renewable energy technology that appears to have the greatest long-term potential nationwide. Three cell technologies are competing to achieve the efficiency and price necessary to succeed in the multibillion dollar utility market:[12] (a) high-concentration systems, (b) single crystal silicon cells, and (c) ribbon silicon or amorphous silicon cells. Each of these devices has at least a chance of achieving the 10-15 percent efficiency and $1-$2/peak watt price that could make it competitive in the utility marketplace.[13] Although each of these technologies faces obstacles that must be overcome to reach these goals, many experts believe that at least one is likely to succeed in the next two or three decades.

Several solar photovoltaic demonstration plants with capacities of about 1 peak MW are currently in operation. The Sacramento (California) Municipal Utility District (SMUD) is operating the largest photovoltaics generating facility owned by a utility.[14] This 1-peak-MW plant began operation in July 1984. SMUD proposes to increase this plant's capacity to 100 peak MW by the year 1993. This expansion would require a large infusion of outside funding, which has not yet been obtained.

Frost and Sullivan, a market research firm, estimates a potential for 6 peak GW of photovoltaic generating capacity by the year 2000,[15] and 100 peak GW by the year 2010, equivalent to approximately 30–40 nuclear power plants. Despite this projection for a large market over the long term, the lack of a major current market for solar photovoltaics at today's prices has led to the departure in the last year of several large companies from the solar photovoltaics business.

Wind is the renewable technology with the largest near-term potential for New York State. Within the next 5–6 years, when new capacity may first be needed in New York State, California's operating experience with windmills of up to 0.5 MW in size should provide New York utilities with a wind generation technology having a relatively low technological risk. If wind-energy generators can be priced competitively with alternative energy sources, there are sites in New York that would make wind a serious option for the state. For the year 2000, EPRI and others have estimated a 20-peak-GW market potential for wind power.[16] In the long term, however, wind has a nationwide potential substantially below that of solar photovoltaics.

The *solar pond* is another promising solar technology, and the only earth-based solar technology that provides the potential for base-load energy supply. Solar ponds are shallow bodies of water that collect solar radiation, store it as heat with a relatively high thermal gradient, and convert this heat into electric power. Thermal convection can be suppressed by dissolving salt in the pond at concentrations that increase with depth. Solar ponds can retain their thermal gradients for long periods of time, and thus can generate electricity when it is needed, not only when the sun is shining. Construction of an economical solar pond requires inexpensive flatland, ample radiation, cooling water, and a low-cost source of brine. One reason for the relatively large land area required for solar pond electric energy generation is that the relatively low temperature differentials (60–80°C) imply very low overall conversion efficiencies (2–3 percent). The cost of utility-scale systems in 1983 ranged from $3,000 to $5,000/kW.

The *central receiver system* (CRS), or "power tower," probably has the least technical, economic, and institutional feasibility of the major solar technologies with potential for electricity generation. The CRS consists of a receiver structure positioned at the top of a large array of sun-tracking mirror, called heliostats, which transmits sunlight onto a central receiver. One drawback of the CRS is that a material-intensive heliostat array about 7,000 square meters in size is required for each megawatt of thermal energy produced. This expensive reflecting array required for current CRS plant designs would make it difficult for such plants to be cost-competitive with fossil-fueled plants. A plant similar in design to the 10-MW Solar One pilot plant installed at Barstow, California, under DOE funding has been estimated to cost about $4,000 per installed peak kW in production quantities.[17]

The *vapor geothermal* energy resource is most available and has been most intensively developed in California;[18] it has a current electric generating capacity of about 1.1 GW. The Pacific Gas and Electric Company plans to add an additional 1.1 GW by 1990. *Hot dry rock geothermal* energy, on the other hand, is widely distributed throughout the nation, including New York State, but as yet is commercially undeveloped. It has been estimated that 1 percent of the hot dry rock energy at depths less than a few miles could produce all the electricity currently consumed in the U.S. At present, however, no one knows whether effective ways can be developed to tap this energy resource.

Other renewable technologies have relatively low near-term potential for providing major new sources of electricity generation, particularly for New York State. *Biomass*, while providing opportunities for using available fuels such as municipal waste, suffers dramatically from the need for large land areas to produce the resource. *Ocean thermal* and other *ocean systems* generally have low potential except for specific geographic areas. *Solar power satellites* could be a major new electric energy source, but the implementation of this technology appears to be at least four decades away.

Nuclear Technologies

The third class of potential new technologies for energy generation includes the nuclear technologies, including *advanced standardized conventional fission plants* and *fusion* or *fusion-fission breeders*. Utility generating plants based on current nuclear technologies can be engineered and built. The question for the next few decades is whether any U.S. utility will buy them, especially if cost-competitive alternative technologies exist. Development of high-availability, inherently safe nuclear reactors that are cost-competitive with coal-fired plants at 200–400 MW electrical capacity could change this picture by the end of this century. The small size would keep the total capital cost of such a plant to a small fraction of the potential purchasing utility's total value and thus could make purchase more likely. Technologies such as the high-termperature gas reactor, which offer the high effluent temperatures needed for industrial process heat, could provide substantial co-generation potential in a relatively small, economically attractive generator. Fusion, especially in some proposed compact high-density versions, may be economically attractive but is not likely to be available for another four or five decades unless substantial unforeseen breakthroughs occur.

Alternatives to New Capacity

A final class of technologies, alternatives to installation of new capacity, may be of interest in an electric utility context. *Energy storage* technologies transfer energy availability from low- to peak-usage periods, and therefore defer the need to build new primary-energy generation capacity. *Pumped hydroelectric* plants, examples of this class of technology, are in operation today, but economies of scale mandate that these plants be quite large (at least 1 GW). *Compressed air* and other storage technologies are being tested and offer the potential of economic storage at sizes an order of magnitude smaller than pumped storage hydroelectric plants. *Demand management* technologies allow scheduling of selected loads during periods of heavy demand and thus also blunt demand peaks and defer the need for new capacity.

Conservation technologies offer a limited but substantial energy savings potential. For example, electric refrigerators currently in use are not highly energy efficient. Refrigerators in current production in Japan provide the same capability as those in current production in the U.S. but consume only one-half the electric energy. Refrigerators now consume more than 7 percent of New York State's electric energy, equivalent to the electricity generated by two average-size nuclear power plants. If New York State utilities provided such energy-efficient refrigerators to all their customers, the need to construct one nuclear plant would be eliminated. The utilities could recover the cost of these refrigerators by charging customers for the electricity their old refrigerators would have used. This approach would require none of the operation and maintenance costs required by the generating and transmission facilities that would be needed if residents retained their current refrigerators or replaced them with currently available less-efficient models.

SUMMARY

Several technologies have the potential to make a contribution to New York State electricity production in the next two decades. Co-generation, in one of its technological forms, clearly has substantial potential. Fuel cells should provide a reasonably priced source of quickly available capacity that is clean and relatively efficient. Depending on fuel price and regulatory requirements, coal-water mixture retrofitting of some current oil-fired generators may be cost effective. Finally, wind energy may, like many innovative fads and contributions, blow in from California and add to New York State's electric energy future.

NOTES

1. Edison Electric Institute, *Yearbook of the Electric Utility: 1980*, cited in *Industrial and Commercial Cogeneration*, Office of Technology Assessment [OTA] (Washington, D.C.: OTA, 1982), p. 4.

2. Office of Technology Assessment, *Industrial and Commercial Cogeneration* (Washington, D.C.: OTA, 1982), p. 10.

3. *Energy Daily*, October 3, 1984 (Washington, D.C.: King Publishing), p. 3.

4. N. Lihach, "Fuel Cells for the '90s," *EPRI Journal* (September 1984):6.

5. H. J. Allison, *Present Status and Future Possibilities for Utility Applications of Fuel Cell Systems*, pp. 121–22, speech presented at Energy Information Dissemination Program, July 9, 1984, Stillwater, OK.

6. Ibid., p. 12.

7. *Energy Daily*, December 1984, p. 1.

8. T. Moore, "Oil's New Rival—Coal Water Slurry for Utility Boilers," *EPRI Journal* (July/August 1984):8.

9. *Synfuels*, October 19, 1984, p. 3.

10. *Synfuels*, February 17, 1984, p. 8.

11. J. B. Tucker, "Solar Power Goes on Line," *High Technology* (August 1984):46.

12. R. R. Perez et al., *Photovoltaics in New York State*, ASRC Publication no. 980, 1984, pp. 2–7.

13. W. Johnson, "Alternate Electrical Sources; Mega Bucks from Megawatts," *Solar Engineering and Contracting* (May/June 1984):38.

14. *Energy Daily*, July 13, 1984, p. 2.

15. Johnson, "Alternate Electrical Sources," p. 38.

16. Congressional Research Service [CRS], *A Perspective on Electric Utility Capacity Planning*, CRS-85, Policy Paper (Washington, D.C.: CRS, 1983).

17. Renewable Energy Institute, *Annual Renewable Energy Review, Progress through 1983*, Alexandria, VA, p. 62.

18. *Energy Daily*, September 26, 1984, p. 2.

PART IV

ENVIRONMENTAL AND HEALTH EFFECTS OF ELECTRIC POWER GENERATION AND USE ⸺

Two major concerns the public has with the electric utility industry are the price of electricity and the environmental and health effects of its generation and transmission. Concerns about price are usually reflected in rate hearings before appropriate regulatory agencies and in the demand for electricity. The latter was treated in Part II; the role of regulatory agencies is the focus of Part VII. Environmental and health effects are reviewed and analyzed in this section.

The four papers in this section explore aspects of the environmental and health effects of the generation and transmission of electric power. The topic is introduced by Senator Robert Stafford (R-Vt.), who provides an overview of the problems and prospects from both public policy and legislative viewpoints. The three subsequent chapters deal with these effects in a more detailed manner and from a more technical perspective. Both fossil fuels and nuclear energy are included in these analyses.

Senator Stafford summarizes the environmental perspective on the generation of electric power using fossil fuels and also nuclear energy. He then introduces the topic of acid rain, outlining what is known and what is not known about its origins and its environmental impacts. Within a public policy framework Senator Stafford considers how to balance the scientific uncertainty about the effects of acid rain with the responsibility of public officials to deal with its economic and environmental risks to many sectors of society. He also asks who should pay the costs of controlling the environmental impact of acid rain—polluters who benefit economically from their own actions, the users of the goods being supplied by polluters, or society in general. The public policy perspectives Senator Stafford provides on the question of acid rain are, of course, relevant to a host of related issues: nuclear energy, the aesthetic insult of high-voltage transmission lines and other kinds of pollution in addition to acid rain.

Harry Hovey pursues these issues with respect to New York State but from a more technical perspective. First, he discusses the effects in New York State of current acid precipitation from sources both within the state and from other regions of the nation and from Canada. He then briefly reviews the history, rationale, and current status of New York State's policy on converting existing oil-burning facilities to coal-fired facilities. Finally, he traces the impacts of these two considerations on the future fossil fuel-combustion regulations prescribed by the New York State legislature. Although each state must control the pollution it creates, without similar actions on the part of other states, no single region will be able to solve its own acid rain problems.

In the third chapter in this section, William Stasiuk takes us deeper into some of the important technical issues associated with measuring the public health effects created by the generation of electric power from fossil fuels and from nuclear fuels. Although the topic is too large to be covered in detail in one chapter, Stasiuk provides an introduction to some of the more important methods of analysis of this relatively complex area of public concern and briefly reviews the available scientific information on these issues.

Although a great deal is known about the impact of electric power generation on public health, there is still a considerable degree of imprecision and uncertainty in our knowledge of these relationships. Whereas Senator Stafford considers this lack of scientific knowledge from a national public policy perspective and Harry Hovey considers it from a state's perspective, William Stasiuk's presentation is primarily from a scientific perspective. For the important analytical issues that he deals with, he shows exactly where our knowledge about the causes and effects of electric power generation on public health is incomplete.

In the final chapter in this section, Alan Crane focuses on nuclear energy and reviews our national experiences with it. After a careful analysis of the pros and cons of nuclear energy, he summarizes the present use and future potential for coal-fired power plants and, briefly, for conservation and alternative sources of energy for generating electricity such as solar, wind, and wood.

The question of which set of options society should pursue is too broad to result in responsible policy recommendations from a single chapter. But Crane's overview is a fitting conclusion to this section because it provides a clear statement of the environmental and health issues involved in choosing among these methods of generating electricity and also provides interested readers with references for works that contain more information about and technical analysis of these issues.

15

Energy and the Environment: The View from Washington ——————

Robert T. Stafford

I am chairman of the Senate Committee on Environment and Public Works, which has jurisdiction over—among other things—matters dealing with the environment and also with regulation of the nuclear electric power industry. For many years I have been an advocate of protection of the environment and also an advocate of strong safety measures in the development and use of nuclear energy.

Over the many years that I have been concerned with the hazards of pollution—particularly to the quality of the air we breathe—I have become convinced that we would all be better off if we were able to eliminate or at least reduce the need for combustion of all sorts.

The smoke from home chimneys, from auto and truck tailpipes, and from smokestacks of factories and power plants is harmful to our lungs, to our eyes, and to the fragile layer of air that envelops our planet. For that reason, among others, I have given my support in the past—and probably will continue to give my support in the future—to federal efforts to encourage methods of providing usable energy from the sun and the winds and the tides.

We have not done a very good job in this nation—nor have others in other nations—of harnessing the energy of the sun (and, thus, the winds and the tides). The odds are that we will not see any significant gains along those lines in the short run. But we should persist, and I imagine we will, in our efforts to encourage greater development of these methods of harnessing energy. They are, for the most part, renewable sources of energy. They also appear to have few unwanted environmental consequences, although experience has taught us that there are probably no human endeavors without any environmental hazards.

179

The Reagan administration has not been particularly supportive of federal government initiatives in this area, and I do not expect it to change in the years ahead, particularly given our federal budget deficits. In short, then, I would anticipate that Congress will continue modest support for development of these alternative methods of producing energy, but I look for no dramatic policy changes.

As our nation has become more aware and more concerned about the environmental and health consequences of the kinds of pollution created by combustion of fossil fuels to produce electricity, I have noted with some surprise the parallel decline of the nuclear power industry. The use of coal and oil to produce electricity has created a broad range of environmental problems, beginning with the mining of coal and drilling of oil and ending with the pollution of our air and water and earth in the generating process. Surely there are many environmental and safety problems that are associated with the production of electric power by use of nuclear fuels. Those include the pollution associated with mining operations; the hazards of the generating process; and our seeming inability to deal with the problem of nuclear waste and spent fuel rods. And I am well aware that the nuclear power industry, while attempting to deal with both the real and the perceived safety issues, was struck a devastating blow by the accelerating costs of nuclear technology and the even more rapidly increasing costs of the capital needed to pay the bills. Still, it seems to me that the industry—the nuclear electric power industry in all of its parts—has to share the blame for lost opportunity along with inflation, public fears, government indecision, and scientific uncertainty, to name just some of the elements that have contributed to the decline of the industry.

In any event, I expect that President Reagan and his administration will continue to support the nuclear power industry. But I do not expect that to be enough to cause the kind of renaissance in nuclear power that is needed to ease significantly the environmental and economic problems faced by electric utilities. I suspect the Congress will not join with the administration in the effort to bring new life to the nuclear power industry until such time as the industry demonstrates clearly that it is ready to meet the public's standards of health and safety. I know that in 1984 members of my committee were unwilling to take any action that might be perceived by the public as a lowering of safety standards regarding nuclear electric power. I do not anticipate any change in that regard in Congress in the near future.

Clearly the nuclear power industry will first have to satisfy the concerns of the American public before it can expect significant new support from the Congress. Those of you who represent that industry or who are closely involved in its dealings are better qualified than I to speculate whether the industry has the ability or even the willingness to undertake that challenge. All of which brings me to my conclusion: unless we find some way to do with less electric power in this country we are going to have to do a better job of protecting our health and environment as we continue to use fossil fuels to produce our electricity.

Which, of course, brings me to the topic of acid rain. As you may know, in addition to being chairman of the Committee on Environment and Public Works, I also represent the state of Vermont in the United States Senate. I am also the principal sponsor of legislation that would mandate reductions in sulfur dioxide emissions, one of the major causes of acid deposition. That acid rain control legislation was originally developed during the 97th Congress in 1982 in the Committee on Environment and Public Works. It was approved in that committee by a vote of 15 to 1 after extensive hearings, analysis, and debate. The Senate did not act on that bill that year, and so I introduced virtually the same bill in the 98th Congress. Once again the Committee on Environment and Public Works gave its overwhelming endorsement to the idea of mandating cuts in sulfur emissions. This time the committee voted for a bigger cut in those emissions and the vote was 16 to 2. But once again the Senate did not act because the House was unable to produce any legislation from its committee of jurisdiction. I can assure you that I will introduce the legislation again in 1985, and I have hopes that we will finally get congressional action on the issue.

Let me outline to you the Senate proposal, which I anticipate will serve as the basis for final congressional action. It is a straightforward approach to the problem. It establishes a program to reduce sulfur dioxide emissions in a 31-state region east of, and adjacent to, the Mississippi River. Emissions would be reduced by 10 million tons annually over a ten-year period, and the costs of those reductions would be borne by the individual polluting facilities. No clean-up strategies are mandated by the legislation. Decisions regarding such strategies would be made by the individual facility. Research would be speeded up to help fill the gaps in our knowledge.

I plan to introduce that legislation early in the 1985 session of the Congress. I anticipate there will be a consensus among the committee members that there will be no need for prolonged hearings, although we may hold some hearings to update scientific evidence that has been developed since previous hearings.

I do not have to repeat here that there are many points of view regarding acid deposition. But perhaps we can agree on some facts: one of the facts is that sulfur dioxide emissions have actually declined since early in the 1970s, a short time after the tough Clean Air Act of 1970s, and dropped to about 24 million tons in 1980. But in 1950 sulfur emissions in this country were about 11 million tons a year—less than half of the total for 1984. In addition, nearly all of the reduction in sulfur emissions from the peak of the early 1970s came from sources other than the electric utility industry. The vast bulk of those reductions came from industrial combustion sources; from commercial, institutional, and residential sources; and from industrial processes.

Now, who contributes the most sulfur dioxide emissions in the 31-state Acid Rain Mitigation Strategy region—the so-called ARMS region—identified in the Senate proposal? Emission inventories show that electric utilities account for nearly three-quarters of the sulfur dioxide emissions in those states east of, and adjacent to, the Mississippi River. Industrial boilers and processes account for less than one-quarter of that pollutant. Cars and trucks emit only about 2 percent of the sulfur dioxide in the East. Those figures undoubtedly explain the utility industry's interest in the pending proposals to mandate reductions in sulfur dioxide emissions.

We have regulated the emissions of sulfur and nitrogen under the Clean Air Act for nearly fourteen years because we know those emissions are hazardous to our health. What we did not realize until relatively recently is that there is another good reason to regulate the oxides of sulfur and nitrogen: once emitted, those pollutants can be transformed into acids. Those acids can, in turn, damage lakes, streams, crops, forests, and soils.

Our scientific community already knows a great deal about this process and will know much more in a few more years. But even a strong advocate of acid rain control like myself will concede there are gaps in our knowledge. We know, for example, that oxides of sulfur and nitrogen are transported in the atmosphere, sometimes several hundreds of miles. But the pathways of transport are less clear. Emissions can be pinpointed with relative ease and great accuracy. But determining their exact destination is much more difficult. Similarly, we have learned that oxides of sulfur and nitrogen can be—and often are—transformed during their travels into acids and acid-forming compounds. The National Academy of Sciences has already provided us with information that indicates the transformation takes place at a linear rate, but we are still uncertain about the exact rate of this transformation. Finally, while we have learned that acid deposition is almost certainly causing serious damage to lakes and rivers, our knowledge of its effects on soils, forests, and crops is only rudimentary.

A year or two ago, a letter I received from an official of the U.S. Environmental Protection Agency highlighted most of those uncertainties and followed with this conclusion: "These questions and others regarding transformation must be better understood before reasonable scientific judgments can be made with any accuracy." The official who wrote that letter demonstrated that there is a vast difference between judgments that are responsible in science and those that are responsible in politics.

When I talk of being politically responsible, I am not talking about pork barrel politics or the selfish protection of vested parochial interests. I am talking about the obligation of an elected public official to make prudent decisions that protect public interests. You have heard me outline some of the uncertainties that exist in the debate over acid rain. Let me add at this point what, in my view, we have learned with a high degree of confidence:

- It is clear that acid deposition is occurring.
- It is equally clear that a major source of acid deposition is the industrialized midwestern region of our country.
- Acid deposition may or may not cause harm when it falls to earth, but when it falls on soil that is sensitive it almost always causes some harm.
- This harm includes reduced productivity of forests and crops. These reductions in productivity may be large.
- Virtually the entire Northeast is sensitive to acid rain, and acid rain regularly falls over this region. In short, acid rain is falling on ground where it can cause harm and that ground includes two entire regions of the U.S.
- In addition to damaging life on or in the ground, acid rain can damage life in and on lakes and streams. This harm is occurring at the present time.

Thus, it is clear to me, the stakes in the acid rain debate are enormous.

Failing to act now places entire regions of our nation at risk and threatens the economic well-being of farmers, foresters, fishermen, and others. Refusal to act because there is a level of uncertainty can be as unwise and foolish as acting too hastily on the basis of too little information. There are many times when it would be irresponsible for a scientist to ignore uncertainty, but equally irresponsible for a public official to give that uncertainty too much weight in deciding public policy. The present constitutes such a time.

There is evidence of this from, for example, William Ruckelshaus, the gifted (former) administrator of the Environmental Protection Agency, who understands the responsibilities of managing risk in our society. Let me quote from his statement to the Senate Committee on Environment and Public Works during his confirmation hearing. "The problems EPA confronts are hard ones. Throughout—from the definition of the problem to its solution—the Agency confronts enormous scientific uncertainty. It often must act before it is clear what the optimum solution would be." Because of that circumstance, it is important that we learn how to accommodate environmental requirements even where the costs are uncomfortable and the benefits are distant.

An additional burden we must carry is the fact that necessary environmental investments will not necessarily deliver results that can be readily measured and identified. The benefits of our environmental investment may be nothing more visible than the failure of elements of the food chain to disappear under the pressure of an ill-defined—or even unidentified—threat. All of which is to say that the protection of our environment often requires investment based on suspicion and speculation. If we wait for absolute knowledge it will likely be obtained too late to avert disaster.

There are those who oppose the suggestion that the U.S. should commit itself to a program to reduce sulfur emissions. But even opponents concede that we will know much more about the subject in another three to five years. And, when pressed further, they will admit that the design and implementation of a sulfur emissions reduction program will take three to five years, even if we begin today. Thus, it is my view that there is no irreconcilable conflict between developing more knowledge and committing ourselves to action. Indeed, the politically responsible course is to take both actions—and to take them as soon as possible.

I am heartened by the fact that the overwhelming majority of the members of my committee have twice approved an acid rain control proposal. There has been other evidence that those of us who want to control acid rain have made some progress. Surely public interest in the issue and public support for acid rain control have been on the increase since the issue was first raised in the Congress in 1977. The reports of the Reagan administration's own task force and of the National Academy of Sciences have strengthened the arguments in favor of effective controls of the pollutants that cause acid deposition. In 1983, for the first time, the House of Representatives had serious acid rain control legislation on its agenda. Surely the House will revisit the issue in 1985.

I am disappointed that the administration still holds the view that more study is needed before we take any action. But I continue to have hopes that the administration will realize that acid rain control would be good for our country, despite complaints by some constituencies to the contrary.

Let me assure you at this point that I am well aware of the enormous economic stakes involved. My proposal has long been the target of those who have charged that a meaningful acid rain control program would dislocate the coal industry; send electric rates out of sight; and generally slow down the nation's economic recovery. Those who labor in the environmental vineyard are familiar with those kinds of arguments. Warnings of economic disaster were sounded by those who opposed the Clean Air Act, the Clean Water Act, the chemical Superfund law, and virtually every other important piece of environmental legislation.

History has demonstrated that past claims of economic disaster that would follow environmental achievement were inflated, at best. At their worst, those claims were patently false. History has also demonstrated that the American public wants clean air and clean water and unspoiled earth—and that Americans are willing to pay a fair price to achieve those goals. And history also suggests that the best way to finance the cost of pollution control is to have that cost paid by those who do the polluting, or who enjoy an economic benefit as the result of uncontrolled pollution.

Many electric customers in the Northeast and in the West where strict environmental controls already limit smokestack pollution are paying the price of pollution control today. It would be unfair to ask those electric customers to pay an additional amount to help meet the cost of pollution control in other parts of the country, since they are already paying for their own cleanup. More important, the debate over who is to pay for acid rain control must not be permitted to delay action on the necessary controls.

In the last session of the Congress, the House of Representatives explored—without success—possible ways of cost-sharing to pay for acid rain control. I anticipate the House will find the Senate proposal—the polluter-pays proposal—worthy of more interest in 1985. The House also explored the concept of a nationwide control program that would include the oxides of nitrogen as well as sulfur. There is obviously some merit in a national approach and I suspect that we in the Senate will keep an eye on any such developments.

If the administration continues its opposition to an acid rain control program, that will make the task more difficult. But it is my view that the Congress, if it gets the chance to vote on the issue, is ready to approve such controls. Senator Robert Byrd of West Virginia, the Democratic minority leader in the U.S. Senate, remains opposed to an acid rain control program because of his perception that such a program would be harmful to the interests of his state. The opposition of Senator Byrd poses another formidable obstacle, but one that can be overcome.

In summary, it is my view that legislation to control acid rain is a concept whose time has come. This nation's acid rain problem will only get worse if we do not act quickly. It is in the national interest to reduce the pollution that causes acid rain: I suggest that it is also in the interest of the electric utility industry to control acid rain. It is my view that the decision makers in the electric utility industry would serve their own interests best by having their industry participate in the task of cleaning up its pollution.

It is my hope—and I anticipate achieving that hope—that the Committee on Environment and Public Works will report a package of Clean Air Act amendments, including a strong acid rain control provision, to the full Senate in the near future. I will urge the Senate to act on the package soon afterwards and I will continue to press my friends in the House of Representatives to act in a like manner. The right thing, of course, is to act now to mandate a significant reduction in the emissions of sulfur dioxide that cause acid rain. It is the right thing for the American people, who want it to happen. And, it is the right thing for those industries who surely must understand at this point in time that, one way or another, the pollution that causes acid rain in this nation will be reduced.

The forces that tend to create an atmosphere of conflict between our energy needs and our environmental concerns are not new in this nation. During the Arab oil embargo of 1973, there was great pressure on the Senate Environment Committee to relax some of the requirements of the Clean Air Act. Among the interests exerting that pressure were automobile manufacturers, oil companies, and electric power companies. We resisted that pressure at that time. It is my view today—as it was in 1973—that it makes no sense for this nation to sacrifice its environment in an effort to respond to a temporary energy problem.

If I may, I would like to recall for you some of my comments of twelve years ago on the issue:*

> We must not be frightened into spending any more of our priceless environment in response to the argument that such action is necessary because of the shortage of energy. Our supplies of clean air and other natural resources are not infinite—and we have already wasted too much of those invaluable resources. Americans require adequate supplies of electricity, heating oil, gasoline, coal and other fuels. They are necessities of our modern life. But, we must also preserve the purity of the air around us. Clean air—air that is needed to sustain life itself—is an even greater necessity than adequate supplies of fuel.

In my view, that was a reasonable and sensible approach to circumstances then. I think it is also a reasonable and sensible approach to circumstances today.

---

*Comments distributed to radio stations in Vermont in 1973 at the time of the Arab oil embargo.

16

Air Pollution Impacts of Fossil Fuel-Fired Electrical Generation

Harry H. Hovey, Jr.

The subject of fossil fuel combustion is about as broad an area for discussion as can be found. Fossil fuels heat our homes, businesses, and factories. They provide electricity to light these as well. They power the vehicles that move our goods and ourselves from place to place. The combustion of fossil fuels pervades every aspect of our society. Similarly, the consequences of burning fossil fuels affect society in many ways. Some of these are clearly apparent to the average person and register through our senses as being undesirable at best or perhaps even a threat to health in some cases. We react negatively to black smoke billowing from a chimney or to the choking effect of being downwind of some noxious combustion exhaust. For readily apparent effects, a legion of state and federal regulations to control emissions from the combustion of fossil fuels have been developed. However, a number of effects exist that are not immediately or readily apparent but that present problems as serious as those for which regulations and standards have been established. This chapter addresses the problems and ramifications associated with the current effect of acid precipitation in New York State; the conversion of existing oil-burning facilities to coal firing; and future fossil fuel combustion requirements. Though not a comprehensive review of all the issues surrounding fossil fuel combustion, this discussion will highlight areas that are currently receiving greatest attention and that are prompting reevaluation of policy and position at the highest levels of government.

The concerns about air pollution from fossil fuel electric-generating stations have changed emphasis over the years. In the early part of the century our objective was to reduce smoke and control total particulate emissions. Then, beginning in the mid 1960s we became concerned with reducing sulfur dioxide. This was followed by proposals for control of nitrogen oxides; currently we are looking at federal regulations related to fine particulates. However, the direct impact of these rather traditional pollutants is not our only concern. Fuels contain heavy metals that can add to the environmental burden. Secondarily formed pollutants have also become an issue—especially sulfates and nitrates that have been related to both inhalable particles and acid deposition.

Current concerns regarding acid deposition and energy have major implications for the future of fossil fuel combustion. New York State has long been concerned about emissions of sulfur dioxide and other contaminants. These concerns now reach beyond our boundaries and have prompted the enactment of the New York State Acid Deposition Control Act and development of a draft environmental impact statement to address what policy New York should have with respect to sulfur emissions in the future.

ACID PRECIPITATION

Millions of tons of acid-causing emissions, particularly sulfur dioxide and nitrogen oxides, are emitted annually throughout the country. Use of fossil fuel introduces substantial quantities of material into the atmosphere which, carried by long-range transport in the movement of air masses, are eventually precipitated on land as acidic compounds. The mechanisms of this deposition may be "wet" (rainfall), "dry" (particulate deposition), or a combination of both. In the northeastern United States and southeastern Canada, rainfall is often more than ten times as acidic as normal rainfall. Rainfall in the Adirondacks is about one hundred times more acidic than normal rainfall. Approximately 200 lakes in the Adirondacks have been identified as being critically acidic—meaning their ability to support fish life is threatened—and about 260 lakes are potentially in similar danger. Although the scientific documentation is incomplete, the evidence of damage to forests is just as alarming. A number of ecologists have expressed profound concern for the forests in Canada and the U.S. Hubert Vogelman has found evidence of significant damage to the forests on Camels Hump, in Vermont. Fifty percent of the spruce trees on Camels Hump have died since 1965. Similarly, significant damage to maple and birch trees was observed during the same period.

Forest damage may be the result of a much more complex problem than just acid deposition. Ozone and other pollutants may play an important role in damaging the trees directly, while acid deposition and heavy metals may be contributing to changes in soil nutrition.

There is alarming information being published about the effect of pollution on forests in Europe as well. Throughout extensive areas of West Germany, East Germany, Poland, and Czechoslovakia, certain forests are in advanced stages of decline. Ecologists in West Germany have estimated that 30 percent of the forests in that country are affected. Even Switzerland is concerned about impacts on its forests. A technical representative from Switzerland visited the Department of Environmental Conservation (DEC) in May 1984 to exchange information on damage to forests.

There are two theories on how acidification actually takes place and at what pace. These have been referred to as the "delayed response" and "direct response" processes. In the delayed response theory, acidification takes place over a long period of time. The accumulation of acid deposition over many years eventually depletes the buffering capacity of the area and watershed to neutralize acid. Once the saturation point is reached, more and more acidity is introduced unneutralized, leading to rapid acidification. This theory infers that as buffering capacity is exhausted, more and more lakes and streams will reach a level too acidic to support fish and other aquatic life.

Effects under the direct response theory are more short-term in nature. According to this theory, a body of water has a given rate at which it can nuetralize acidity. Equilibrium between acid inputs and outputs is established rapidly. This theory infers that at present emission rates, no further acidification will occur unless there is an increase in acidic deposition.

Insufficient data exist to prove or disprove either hypothesis, but a group under the direction of Stephen Norton at the Department of Geological Sciences, University of Maine, has found indications that acidic lakes in New York and New England are becoming increasingly acidic even though sulfate flowing into them has not increased for several years. Norton's group took core samples from the bottoms of ten lakes of varying acidity in the Adirondacks and thirty lakes in New England. Radioactive dating techniques allowed Norton to pinpoint the time when each slice in the core sample was deposited. For every lake of pH 5.5 or below, deposits of calcium, which neutralizes acids in water, started falling sharply in about 1950 to their current level, which is only about one-third of what it was 100 years ago, according to Norton. The Environmental Protection Agency (EPA) plans to do more work to confirm these findings.

Although a substantial amount of data has been gathered on acid deposition in the Adirondacks and elsewhere (Fig. 16.1), the mechanisms involved are complex and not easily relegated to defined processes or established cause and effect relationships. In short, the amounts of sulfur

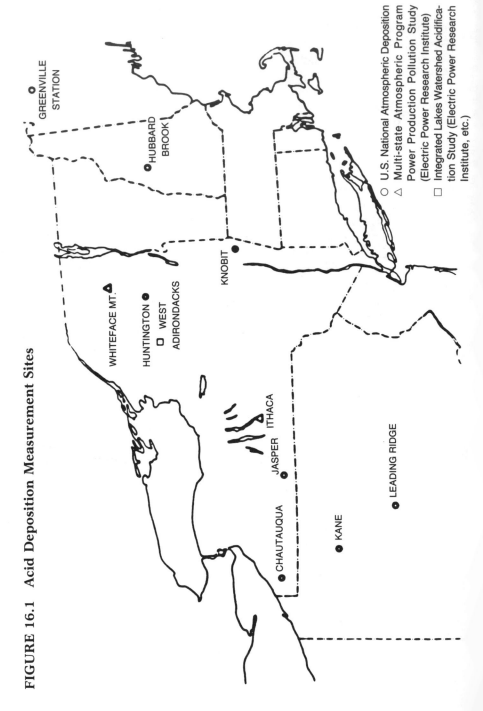

FIGURE 16.1 Acid Deposition Measurement Sites

○ U.S. National Atmospheric Deposition
△ Multi-state Atmospheric Program
   Power Production Pollution Study
   (Electric Power Research Institute)
□ Integrated Lakes Watershed Acidifica-
   tion Study (Electric Power Research
   Institute, etc.)

GREENVILLE
STATION

HUBBARD
BROOK

WHITEFACE MT.

HUNTINGTON
□ WEST
ADIRONDACKS

KNOBIT

ITHACA

JASPER

CHAUTAUQUA

KANE

LEADING RIDGE

dioxide emitted to the atmosphere are well established, and the increase in acidity in sensitive areas has been documented. Sophisticated computer simulations have been developed to estimate the deposition likely to be caused by a given emission of sulfur dioxide. However, data are lacking that directly attribute cause (acid emissions) to effect (increased acidity, forest deterioration, and the decline of fish populations in sensitive areas). What we can determine is that there is a very high probability that the emissions of large quantities of sulfur dioxide and other acidic components from sources in the Midwest are linked to the demise of numerous lakes in the Adirondacks.

Use of fossil fuels has increased substantially in the last 20 years, locally, nationally, and worldwide. In the same period, New York has implemented stringent regulatory controls for sulfur dioxide emissions. Since 1968, New York has reduced its sulfur dioxide emissions by approximately 50 percent, or 1 million tons. This was accomplished during a period of substantial growth in the electric power generation industry. Other areas of the country do not have an equivalent track record (Fig. 16.2). The amounts of sulfur dioxide emitted by power plants in the midwestern states are considerably greater than those in emissions from New York and neighboring states (Table 16.1). The Midwest has long been characterized by large power plants with very tall stacks for dispersion of contaminants from burning high sulfur fuels (3–5 percent sulfur coal). Violations of ambient air quality standards in these states have been largely avoided by the use of the tall stacks to disperse the contaminants, but as a result these emissions have been exported to downwind states, with the subsequent effects we are witnessing today in the Adirondacks.

There has been little activity to date at the federal level to mitigate these effects. The issues are complicated and involve the interests of numerous parties. Accomplishing significant changes in current emissions will require substantial capital investment in control equipment for removal of sulfur dioxide from combustion effluents, or increased fuel costs for lower sulfur fuels or fuel desulfurization processes. All of these solutions translate into higher electric rates for users in the states affected. Although numerous proposals have been advanced in Congress and elsewhere, no approach has yet been put forth that would represent a satisfactory compromise for the parties involved and still effect a meaningful reduction in emissions of sulfur dioxide. Mario Cuomo, governor of New York, in testimony before the House Subcommittee on Health and the Environment, advocated a 12-million-ton reduction in sulfur dioxide emissions by the year 1995. This is the same goal called for by the National Academy of Sciences. He further recommended imposing a tax based on emissions of sulfur dioxide and nitrogen oxides, rather than imposing a tax based on nonnuclear electricity generation. This would place

FIGURE 16.2 Utility Sulfur Dioxide Emissions by Region (Tons per Year, 1982)

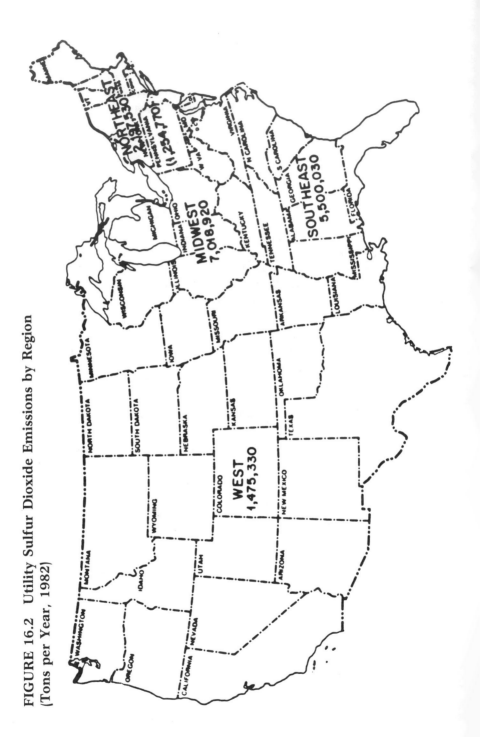

the burden for reducing pollution on the polluter, where it properly belongs. Otherwise, monies paid by New Yorkers to reduce acid rain will be spent largely in other states, raising the rates New Yorkers must pay and subsidizing reductions by others. Needless to say, this approach has not been received favorably by states whose emissions would cause them significant expense.

Because New Yorkers are so directly and measurably affected by acid deposition, they are not allowed the luxury of a long, leisurely debate over the sources of acid deposition, whether these can be diminished or controlled, or what measures should be taken to accomplish this. An immediate review of New York's current sulfur policy is necessary and appropriate.

On August 6, 1984, Governor Cuomo signed into law the State Acid Deposition Control Act (ADCA) which mandates reductions in sulfur dioxide emissions by 1988. In general, this act requires the DEC to identify sensitive receptor areas, establish environmental threshold values for such areas, and promulgate a program to achieve appropriate reductions in sulfates at the sensitive areas. This law was passed coincidentally with the DEC's development of a new sulfur policy. Therefore the draft environmental impact statement that had been prepared was used to fulfill many of the requirements in the ADCA.

Hearings have been held on the draft environmental impact statement in order to determine practical alternatives aimed at reducing sulfur dioxide emissions in New York State without causing unacceptable social, economic, and environmental effects. When the final environmental impact statement is prepared, it will provide a basis for promulgating regulations to effect the changes that are determined to be appropriate. Without accompanying national action, New York cannot expect to reverse or offset the effects being inflicted by upwind states. It is, however, a start and represents an effort to minimize New York's contribution to the problem. Just as questions of interstate commerce have been resolved in the past, the federal government must accept its responsibility for protecting individual states from interstate pollution.

CONVERSION FROM OIL TO COAL

The energy crisis of 1974 made New York State's dependence on foreign oil and the subsequent vulnerability that this entails dramatically obvious. In response, plans were advanced by the State Energy Office and electric utilities in New York State to shift a portion of the electricity produced from combustion of oil to combustion of coal. To accomplish this goal, given the harsh realities of stable or diminishing electrical de-

TABLE 16.1 Utility Sulfur Dioxide Emissions by State (Tons per Year, 1982)

NORTHEAST			MIDWEST		
Connecticut	45,590		Illinois	1,065,200	
Maine	12,450		Indiana	1,313,770	
Massachusetts	259,020		Iowa	196,790	
New Hampshire	60,990		Michigan	593,620	
Rhode Island	2,970		Minnesota	137,250	
Vermont	370		Missouri	1,116,800	
New Jersey	95,670		Ohio	2,179,590	
New York	465,700		Wisconsin	415,900	
Pennsylvania	1,254,770			7,018,920	SUBTOTAL
	2,197,530	SUBTOTAL			
SOUTHEAST			WEST		
Alabama	416,330		Arizona	104,880	
Arkansas	56,860		California	18,050	

Delaware	56,500		Colorado	83,660
District of Columbia	690		Idaho	0
Florida	651,300		Kansas	156,200
Georgia	823,510		Montana	16,970
Kentucky	886,600		Nebraska	40,300
Louisiana	32,820		Nevada	59,690
Maryland	207,130		North Dakota	105,080
Mississippi	967,020		New Mexico	104,540
North Carolina	418,470		Oklahoma	82,490
South Carolina	196,030		Oregon	3,870
Tennessee	644,900		South Dakota	39,910
Virginia	124,740		Texas	449,070
West Virginia	888,130		Utah	22,750
			Washington	56,970
			Wyoming	130,920
SUBTOTAL	5,500,030		SUBTOTAL	1,475,330

Source: Calculations from U.S. Department of Energy Publication DOE/EIA-0191 (82) and printout on actual fuel use for 1982.

mand, specific plants were identified that had previously burned coal and could be considered viable candidates for reconversion from burning oil to burning coal.

There are many problems associated with conversion of these plants to burning coal. In addition to the basic requirement for showing compliance with ambient air quality standards and emission limits, there are numerous potentially offensive activities intrinsic to the operation of any coal-burning facility. Coal must be delivered by water or rail, which entails noise and dust. The stockpile needed cannot be considered an aesthetic improvement to the landscape, and leachate from the coal pile has the potential for contaminating groundwater unless precautions are taken. The operation of the facility itself (coal handling, vehicular traffic, and increased fan capacity for emission control) is a potential source of noise and dust. Even at the best managed plants, the resultant waste streams from burning coal are a major concern. The collection of flyash from particulate control equipment requires disposal at a secure site to preclude leaching and contamination of adjacent groundwater. If scrubbers are utilized, the contamination problem is magnified many times, both in terms of the volume of waste produced and the precautions required for disposal. Conversion to coal requires addressing issues such as toxic trace element emissions associated with fine particulate matter and increased nitrogen oxide emissions. For each of the proposed conversions, these and numerous other factors were evaluated in reaching a final decision. As a condition of each approval, particulates were limited to 0.03 pounds of particulate per million BTUs of heat input, which is equivalent to what is required for new electric generating plants. This limit can be obtained only with equipment and systems capable of removing a substantial percentage (more than 90 percent) of micron-size particulate material (i.e., particle sizes less than 5 microns). Compliance, in turn, removes a substantial number of associated trace metals having an affinity for such particles. A requirement was also imposed to identify and, where possible, implement measures for controlling emissions of nitrogen oxide. Each conversion application was unique and required individual consideration.

The first approval was granted to Rochester Gas and Electric (RG&E) for units 7 and 8 of their Beebee Generating Station. After approval was granted, RG&E decided not to convert because of economic factors. Although the sulfur content of the coal that was planned for use at Beebee was within the limits set by the DEC regulations for coal, conversions would have meant an increase in emissions because the DEC's State Implementation Plan (SIP) allows coal to have a higher sulfur content than it allows oil. To prevent violations of the ambient air quality standards, a limit was imposed on the total amount of sulfur dioxide emissions per hour rather than requiring coal to have a lower sulfur content.

The second approval was granted to Orange and Rockland Utilities, allowing them to emit 1.0 pounds of sulfur dioxide per million BTUs from both units 4 and 5 of their Lovett plant. Approval was granted to emit 1.5 pounds of sulfur dioxide per million BTUs if only one unit were operated at a time. A scrubber option with similar emissions was also approved. These approvals were granted on April 13, 1982, by the DEC. Since the emissions were more than would have otherwise been required under state regulations, an SIP revision request was prepared showing compliance with all standards and sent to the EPA on June 7, 1982. To date, the EPA has neither approved nor disapproved the request. Approval is anticipated to be forthcoming shortly.

The third approval was granted to Consolidated Edison (ConEd) of New York on September 14, 1983, for coal use at their Arthur Kill Station units 2 and 3 and their Ravenswood Station unit 3. The approval requires the use of scrubbers at both facilities. This decision was made by the DEC commissioner, Henry Williams, based on the finding that reductions achieved by scrubbing could be financed out of fuel cost savings. ConEd has subsequently requested reconsideration of this decision and been denied. They have also initiated legal action against the DEC to overturn the decision. The matter is still before the courts and an early resolution is not expected. ConEd maintains that costs associated with scrubbing preclude their conversion of the units to coal. Also, although disposal measures were identified in the public hearing process, the disposal of scrubber sludge creates substantially more problems for ConEd than does ash disposal. Since emissions using scrubbers would be within the limits required by regulation, an SIP revision request was not needed.

The Long Island Lighting Company (LILCO) submitted an application for units 3 and 4 of its Port Jefferson plant on April 28, 1982. Hearings were begun and testimony presented by all parties. During the course of these hearings, legislation was enacted prohibiting future landfills on Long Island. Although the LILCO interprets this legislation as not applying to the Port Jefferson coal conversion, the DEC staff feels that it does. As such, the hearings have been adjourned pending resolution of this critical issue. If the LILCO cannot deposit ash in landfills on Long Island, the conversion of Port Jefferson to coal will probably be precluded.

The last coal conversion application was submitted by the Central Hudson Gas and Electric Corporation for units 3 and 4 of its Danskammer plant on April 20, 1983. After a number of discussions to resolve deficiencies in the application and outstanding questions, the application was declared complete on December 15, 1983. Public hearings commenced on February 23, 1984, and were completed in the spring of 1985. The commission is expected to make its decision in the summer of 1985. The application is complex in that it involves an oil-fired electric generating facility in proximity to the proposed coal conversion, which has impacts at coincident points. Emissions from this facility must be considered in the demonstration of compliance with ambient air quality standards. Violations of the sulfur dioxide National Ambient Air Quality Standard were recently measured in the vicinity of the Central Hudson plants. Diffusion analyses indicated that the Roseton plant was the primary source of sulfur dioxide and that a reduction in sulfur content of oil burned there from 2.0 percent to 1.5 percent would provide for compliance with ambient standards in conjunction with the coal conversion at Danskammer.

Although other plants and units were identified in the New York State Energy Master Plan, these have, for one reason or another, been found to be unlikely candidates for coal conversion. No applications, other than those noted, have been submitted for coal conversion.

OUTLOOK FOR FUTURE REQUIREMENTS

Because of the acid deposition situation in New York, fossil fuel combustion is likely, if not certain, to be subject to more stringent requirements to achieve reductions in emissions of sulfur dioxide. In the past, the central thrust of most air pollution control programs was to achieve and maintain compliance with ambient air quality standards. To this end, strategies were adopted that were less concerned with the absolute amounts of a contaminant that was emitted than with the ground level impact of that contaminant. A representative case concerns the issuance by the DEC of special fuel-use limitations. A provision exists in our regulations for use of higher-sulfur-content fuel than would otherwise be permitted, provided the source owner shows to the commissioner's satisfaction that all standards will be met. It has become apparent that such actions, however acceptable they may be by themselves, must be viewed in an overall context that includes the current level of concern and scrutiny. To this end, the DEC reevaluated its total approach to limiting sulfur emissions. Much closer attention was paid to cumulative emissions and the criteria used to limit them. The DEC's assumption of responsibility for administering the federal Prevention of Significant Deterioration (PSD) regulations was one step in this direction.[1] In the future, assessment of acceptability will be based not only on compliance with emission limits and ambient air quality standards but also on an evaluation of the cumulative effects of multiple sources on acid deposition. We cannot view individual actions as having no relationships to larger issues. The exact method to be employed will be determined by the outcome of hearings on the draft environmental impact statement, but it is safe to say that emission requirements will never again be as they were.

The ADCA signaled a new set of concerns for fuel users, especially the electrical utilities. In the legislative findings, the legislature determined that acid deposition was occurring throughout New York State, that the sources of acid deposition were from both inside New York State and outside New York State, and that, although federal legislation was needed, New York State should do its fair share.

A key to the legislation is the term "sensitive receptor areas." These are regions of the state encompassing significant geographical land areas, which the DEC determines to be susceptible to impacts of acid deposition. The definition of a sensitive receptor area says that no sensitive receptor area can be so small as to be contained within any one county of the state; so a sensitive receptor area has to be relatively large and has to cross the borders of more than one county.

The determination of sensitive receptor areas is based on a number of factors: a) geological information—the DEC is concerned about both the bedrock and the overlying soil structure; b) the presence of sensitive plants and animals—if there are no sensitive materials, plants, or animals in an area, then although all the other factors may prevail, an area would not be considered a sensitive receptor area; c) information from existing reports and data—the DEC staff in Fish and Wildlife have done a significant number of reports on the impacts on the lakes in the Adirondacks, and much information exists concerning the water quality and the quality of the fish life in these various lakes.

Another key definition in the law is the term "environmental threshold value." This is defined as a deposition rate at which no significant effects of acid deposition would occur in kilograms per hectare, which is almost equal to pounds per acre. For example, 20 kilograms (kg) per hectare would be roughly 18 pounds per acre. As part of a recent environmental impact statement, the DEC has made the statement that it believes that 20 kg per hectare of wet deposition is, in fact, the environmental threshold value that should be used in New York's sensitive receptor areas. The sensitive receptor areas in New York are being defined by this environmental impact statement; obviously the most sensitive area, the most critical area, includes the western Adirondacks, and 20 kilograms per hectare is the deposition number that the DEC has proposed for that area. Public hearings on the environmental impact statement were held in September 1984 and in the beginning of October. After taking into consideration all of the comments and following preparation of a final environmental impact statement, it was determined that no significant change should be made in the environmental threshold value and, therefore, 20 kilograms per hectare was retained.

Another important factor in the legislation is a final control target. This legislation is significantly different from proposals discussed at the federal level. New York is not talking about a 20 percent, 30 percent, or "X million tons" reduction in sulfur dioxide but, rather, about a reduction in sulfate deposition. The final control target is calculated by taking New York State's contribution to the sensitive receptor areas, multiplying that by the difference between the wet deposition and the environmental threshold value, and dividing the difference by the wet deposition. Then, using that same form of the formula for emissions and deposition, calculated by a mathematical model, we are able to determine what the reduction should be from various areas of the state and facilities within the state.

When the legislature passed the ADCA, they decided that it would not be appropriate for New York State to go all the way to the final control target initially, and that the first step should be what they called an "interim control target." The interim control target is approximately 40 percent of the final control target. This is the program that must be in place by January 1, 1986, and sources have to be in compliance by January 1, 1988.

The law requires public hearings, which were held in early March 1985, on the preliminary control program and on the regulations that are needed to meet the interim control target in the spring of 1985. These hearings addressed the form of the regulation and the various concepts that will be in the regulation; later in 1985 public hearings will be held on the regulation itself. That still leaves time to meet the deadlines in the law.

This legislation applies primarily to major steam electric generating facilities that are larger than 50,000 kilowatts (kW) or stationary sources that emit more than 100 tons per year of sulfur dioxide. That covers a significant number of sources. The law also covers stationary sources that emit more than 100 tons per year of these nitrogen oxides and requires the adoption of controls for new sources of nitrogen oxides, but these controls will not be adopted until 1987. The department will have to prepare an environmental impact statement, which is approximately a two-year process.

The other feature of the law is that by January 1, 1991, the DEC must go back to the governor and the legislature with a final control target. They will then decide whether or not New York State will make the next step, based on what has happened at the federal level, or what other states have done if no federal program exists.

The regulatory requirements to meet New York's share of a 20 kg per hectare of wet deposition environmental threshold value in the sensitive areas would be equivalent to a limit of the sulfur content in oil of 1.5 percent and in coal of 1.7 pounds per million BTUs. The current regulation for upstate New York sources is 2.0 percent sulfur in oil and 1.9

pounds of sulfur in coal. There are certain areas where the numbers are lower: for example, in the Buffalo area 1.4 pounds of sulfur in coal, and in New York City, 0.3 percent sulfur in oil. These areas would stay at the lower number.

A second concept in the regulation is the idea of offsets or tradeoffs. The law provides for offsets—that is, offsets against the deposition, rather than against ton for ton of emissions. Therefore a source relatively close to the western Adirondacks that has a strong influence on that area could, by instituting additional controls, offset a significant amount of emissions from sources that are farther away, because the mathematical models that are used all show that deposition is related to distance as well as to source strength; and the farther away the source, the less impact it has on the location.

With regard to Canada, in the western Adirondacks Canadian sources are responsible for almost 30 percent of the deposition. So if Canada moves forward on a 50 percent reduction program, the U.S. may get as much as a 15 percent reduction in deposition just from Canadian sources. New York State is responsible for about 18 percent of the deposition in the western Adirondacks; if the state program moves forward as planned, a 6-7 percent reduction for the initial interim target should result. Those two numbers together might add up to a large enough change in deposition that we would be able both to measure the impacts and to determine the future control program.

Finally, the concept of mothballing a plant was discussed at one of the public meetings on the ADCA. Some utilities in New York State may currently have some excess capacity, and the question was raised, would it be viable to mothball a plant and take credit for it against deposition so that the company would not have to reduce other facilities as much? This concept is being considered in the development of the regulation. However, if the mothballed plant were to come back on-line, one of two things would happen: either it would have to meet federal performance standards for new sources, or, if it were to come on-line as an existing plant that did not have to meet federal new source performance standards, a reduction in deposition equal to that which was previously offset would have to occur. One of the ideas suggested to bring about such a reduction was a scrubber on an existing plant at the time that the first plant is brought back into operation.

SUMMARY

Fossil fuel combustion is rapidly creating environmental constraints that will change past usage patterns for fossil fuel. We have encouraged and welcomed the improvement in air quality resulting from the recent increased use of natural gas. However, we cannot depend on this remedy for the long term.

The existing acid deposition problems being experienced by New Yorkers and others can only be arrested by achieving reductions in acid precursors. New York is taking steps to see that, in addition to the major environmental advances already realized, emissions of sulfur dioxide in the state are reduced to the maximum extent practicable. It remains for other states to recognize their responsibilities and obligations toward their neighbors and to effect similar remedies. Without a comprehensive national effort, there is little likelihood of voluntary contribution to such an effort. The DEC believes the time to act is now and has called upon the Congress and the Reagan administration to do likewise. We are hopeful their response will be timely and effective.

NOTE

1. The PSD regulations require that the cumulative impact of new sources be evaluated annually.

17

Some Public Health Concerns Associated with Electrical Energy Generation

William N. Stasiuk

INTRODUCTION

Public health concerns associated with electric power generation are wide ranging and have generally evolved from observations of cause and effect. Epidemiology and risk assessment techniques are the scientific tools used to attempt to quantify these concerns.[1]

Any thorough consideration of the public health aspects of electric power generation must deal with the entire fuel cycle, including mining and processing raw materials, and disposal of waste materials, as well as the impacts associated with the generation of energy. Any societal choice from an array of energy options, if based on a comparison of risk, will by necessity become a political one that must evaluate and account for public acceptability of the risk. Public acceptability, in turn, takes into consideration not only the numerical description of the risk but also the nature of the risk.

From a practical point of view, nuclear and fossil fuels are the two options for future electric power generation in New York State. Fossil fuel is essentially a coal option. The risks associated with either fossil or nuclear fuel are often different and vary with:

- the certainty of cause and effect;
- the probability of the cause;
- the nature of the effect;
- public perception of the risk.

The following are examples of these differences. The risks ascribed to fossil fuels are somewhat imprecise because the quantification of the relationship between dose and effect for air emissions is not clearly defined and remains controversial. On the other hand, the cause and effect relationship from ionizing radiation is well quantified and less controversial. However, the public acceptability of risk from nuclear reactor accidents, regardless of how low the probability, or of risk from the long-term nuclear waste disposal options, remains a major issue.

HEALTH EFFECTS FROM THE USE OF FOSSIL FUELS

The public health concerns associated with electric power generation from the combustion of fossil fuels include occupational safety and risks associated with exposure to gases and aerosols in occupational and environmental settings (see Table 17.1).

There are clear risks associated with working in coal mines. Mortality studies of coal miners indicate that miners die more often than other people from respiratory diseases, stomach and lung cancer, and hypertension. In a significant number of cases, pneumoconiosis and bronchitis were underlying causes of death. The environment around mining towns and refineries typically contains higher than average levels of aer-

TABLE 17.1 Risk Areas

Fossil	Nuclear
Mining • injury • occupational disease • community pollution	Mining • occupational • community exposure
Processing/refining	Processing • environmental • occupational
Transport	
Electrical generation • air pollution	Electrical generation • air pollution • occupational exposure • catastrophic accidents
	Transportation
	Decommissioning
	Ultimate waste disposal

osols and/or hydrocarbons. In addition, though safer now than in earlier times, mining is still dangerous. In 1977, 139 fatalities were recorded among coal workers. From 1970 through 1977, 420,000 coal miners or their widows were awarded federal Black Lung Compensation because of disability or death as a result of pneumoconiosis. The number of other work-related illnesses is impossible to estimate. (For an in-depth discussion of workplace and community implications of coal mining, see the Office of Technology Assessment document, "The Direct Use of Coal."[2])

In the nonoccupational setting, the primary concern associated with use of fossil fuels pertains to combustion-related air pollution. The increases in mortality and morbidity associated with air pollution at the very high levels present in European and American cities as late as the 1950s and 1960s are well documented. As a result, in both Europe and the U.S., control measures have been taken that have markedly lowered community exposure to pollutants such as sulfur oxides and particulate matter.

Despite the historical evidence of cause and effect, the health implications of current air quality and its spatial and temporal variation are quite unclear, as are potential changes in mortality and morbidity were pollution levels to increase or decrease.

The comprehensive Community Health and Environment Surveillance System (CHESS) studies remain controversial,[3] as does the work of L. B. Lave and E. P. Seskin,[4] two often quoted sources for threshold levels (i.e., levels below which there are no adverse human health effects) and the quantification of the relationship between dose and effect. All of these studies are frequently used in estimating risks of increased use of fossil fuels.

One particularly comprehensive critique of these and other air pollution epidemiology studies appeared as an entire issue of the November 1979 American Journal of Epidemiology and was prepared by the epidemiologists most involved in studying British health problems related to air pollution.[5] In the issue of that journal immediately following, C. M. Shy presents an articulate rebuttal of these British views from the American outlook, particularly emphasizing the conservative public health requirements of the Federal Clean Air Act.[6] Suffice it to say that the issues surrounding thresholds and relationships between dose and effect remain controversial.

This uncertainty is primarily a result of the complexity of determining cause and effect via the science of epidemiology, which attempts to estimate the statistical relationship between the dependent health variable and the independent variable of pollution exposure. Statistical corrections must be made for other influential variables such as age, sex, smoking habits, occupation, socioeconomic factors, and other exposures, especially those occurring in indoor environments. The last factor cannot be overemphasized. Questions have also been raised about the validity of using data from ambient monitoring stations as indexes of environmental exposure.

Unfortunately, this controversy is likely to remain unresolved. There is little evidence that significant resources or enthusiastic efforts are being directed toward improving the epidemiology of adverse health effects attributed to air pollution that results from the combustion of fossil fuels.

NUCLEAR

As with fossil fuels, the public's health concerns pertain to the entire nuclear fuel cycle. Many authors have attempted to quantify those risks.[7] These risks include but are not limited to the risk areas listed in Table 17.1.

I would like to focus on the event of most concern to the public, namely the likelihood and consequences of a significant accident involving the nuclear fuel within a reactor core; I will also discuss the emergency planning zones for off-site disaster preparedness in the vicinity of nuclear power plants. The methodology used to analyze the public health aspects of an accident involving reactor core damage is called "probabilistic risk assessment," or PRA. In order to introduce the concept of PRA, some discussion of safety design considerations is appropriate.

According to the "Reactor Safety Study,"[8] the safety design approach for nuclear power plants consists of three levels of safety, involving (a) the design for safety in normal operation, providing tolerances for system malfunctions; (b) the assumption that incidents will occur despite (a), and the inclusion of safety systems in the facility to minimize damage and protect the public; and (c) the provision of additional safety systems to protect the public based on the analysis of very unlikely accidents.

The safety design approach has also been described as the so-called defense-in-depth approach in which multiple physical barriers prevent the escape of radioactivity to the environment. The multiple barriers are the fuel cladding, the reactor coolant system, and the containment structure itself. In addition, several systems are designed as safeguards to provide protection during abnormal events. These systems include, among others, emergency core cooling, standby liquid control, and ultimate cooling water connection.

Although nuclear plants are designed to contain and correct anticipated operational failures, it is conceivable that unanticipated multiple failures of the various defense-in-depth mechanisms may occur. In Nu clear Regulatory Commission (NRC) terminology, these are called "class 9" accidents or "beyond design basis" accidents. Because severe accidents can only occur if there is damage to the reactor core, all class 9 accidents are considered "core melt" accidents. Any significant accident will involve some core melting and subsequent release of radiation.

PRA is the methodology for assessing the probability of occurrence of the various core melt accidents as well as the consequences of these accidents. The PRA methodology, as originally presented in the "Reactor Safety Study," identifies accident sequences using "event trees," determines the size of the radioactive release (source term), and analyzes the impact on the environment using meteorological data from the site. Actual incidents described in Licensee Event Reports[9] can be analyzed by PRA methods[10] and in turn can provide additional data for a PRA.

The PRA process yields a set of possible accidents, each with its own probability of occurrence and its own source term (the amount and timing of the radioactive release). Since the consequences of any release will vary depending upon the weather conditions, each source term is modeled for the various possible weather conditions. The result is an estimate of the probability of exceeding given amounts of radiation at any distance from the plant. These results can be displayed for each accident sequence or aggregated and displayed for all core melt accidents.

Commonly, the results are expressed in terms of conditional probability. They are obtained by the following process:

1. Quantify the fission product release characteristics for each accident;
2. select an appropriate atmospheric dispersion model;
3. use a representative (e.g., one-year) data base of hourly weather;
4. randomly select a representative number (one or several hundred) of daily weather data from a given year's set of data;
5. run the model using the source term of interest for each meteorological period. This yields doses for various distances for each run;
6. compute for each source term a probability of exceeding the dose limit for each distance (note that the probability will come from a consideration of where the data fall in the data set);
7. pool the data at each distance interval by multiplying the probability of exceeding a dose by the probability of occurrence of that accident (release frequency) and summing over all accidents.

This methodology yields a family of curves that describes the conditional probability of exceeding various doses of radiation at any distance. In order to find the probability of exceeding a given dose at a given distance, the conditional probability must be multiplied by the probability of a core melt accident.

This methodology was used to develop the basis for off-site emergency preparedness. In order to introduce this basis, we will first discuss what is known about the public health effects of exposure to radiation and the concept of protective action guides.

Health Effects

The biological changes caused by exposure to radiation may cause harmful health effects. The common unit that refers to radiation dose is the *rem* (roentgen equivalent man). The roentgen unit describes the quantity of ionization produced by ionizing radiation in air. The rem represents the amount of ionizing radiation that produces a biological effect in human tissue equivalent to the effect of a standard type of radiation (such as X-rays). The average person in the U.S. receives about 0.18 rem a year of which 0.11 rem is naturally occurring radiation; the remainder results primarily from diagnostic X-rays. Health effects can be separated into both acute health effects and delayed health effects.

For short-term effects, at exposures of 50 rems or less, ordinary tests show no indication of injury. At 100 rems, some individuals exhibit clinically measurable changes in blood cell count. At 200 rems, most people will show definite signs of injury, and those most sensitive to radiation could die from infection in the absence of medical treatment. At 450 rems, half those exposed will succumb within thirty days unless treated. At 800–1,000 rems, a fatal outcome is virtually certain without treatment and therapy.

Delayed effects, including cancer and genetic changes, may result from a previous short-term exposure or from chronic exposure to low doses over a period of years. The National Academy of Sciences estimates the increase in lifetime risk of cancer mortality from a single whole body absorbed dose of 10 rems to be from 0.5 to 1.4 percent of the naturally occurring cancer mortality.[11] This estimate means that if 1 million people were each exposed to a dose of 1 rem (i.e., 1 million person-rems), the expected number of cancer deaths would be between 77 and 226 (best estimate, 148) in addition to the number of cancer deaths (163,800) normally expected in those 1 million individuals. One can calculate in a similar manner the expected impact of various exposures to different numbers of people.

Dr. Edward P. Radford, chairman of the Biological Effects of Ionizing Radiation (BEIR) III Committee, points out that new estimates of the dose received by Japanese atomic bomb victims require a reassessment of the relationship between dose and effect. He believes a figure of 4,000 excess cancer deaths for every 1 million persons exposed to 10 rems (i.e., 10 million person-rems) is more appropriate than the best estimate of 1,480 based on the BEIR III range of the expected number of additional cancer deaths of from 766 to 2,255.

Protective Action Guides

Protective action guides (PAGs) were developed to assist decision makers in responding to radiological emergencies. As described by the joint Federal Nuclear Regulatory Commission/Environmental Protection Agency (NRC/EPA) task force on emergency planning:

> The concept of PAGs was introduced to radiological emergency response planning to assist public health and other governmental authorities in deciding how much of a radiation hazard in the environment constitutes a basis for initiating emergency protective actions. The PAGs are expressed in units of radiation dose (rem) and represent trigger or initiation levels, which warrant preselected protective actions for the public if the projected (future) dose received by an individual in the absence of a protective action exceeds the PAG. PAGs are defined or definable for all pathways of radiation exposure to man and are proposed as guidance to be used as a basis for taking action to minimize the impact on individuals.
>
> The nature of PAGs is such that they cannot be used to assure that a given exposure to individuals in the population is prevented. In any particular response situation, a range of doses will be projected, principally depending on the distance from the point of the radioactive release. Some of these projected doses may be well in excess of the PAG levels and clearly warrant the initiation of any feasible protective actions. This does not mean, however, that doses above PAG levels can be prevented, or that emergency response plans should have as their objective preventing doses above PAG levels. Furthermore, PAGs represent only trigger levels and are not intended to represent acceptable dose levels. PAGs are tools to be used as a decision aid in the actual response situation.[12]

The PAGs for whole body exposure to airborne radioactive materials are listed in Table 17.2.

TABLE 17.2   Protective Action Guides for Whole-Body Exposure to Airborne Radioactive Materials

Population at Risk	Projected Whole Body Gamma Dose (rem)
General population	1–5[a]
Emergency workers	25
Life-saving activities	75

[a]When ranges are shown, the lowest value should be used if there are no major local constraints in providing protection at that level. Local constraints may make lower values impractical to use, but in no case should the higher value be exceeded in determining the need for protective action.

Planning Basis for Off-Site Preparedness

Figure 17.1 is derived from the "Reactor Safety Study" and includes all accident release categories for both pressurized and boiling water reactor accidents. Based on the conditional probability of dose versus distance, there is a 30 percent chance of exceeding the PAGs at ten miles from a plant in the event of a core melt accident. When coupled with the probability of such an accident, the probability of exceeding PAG doses at ten miles is 1.5×10^{-15} per reactor year. This analysis was apparently a major factor in development of a ten-mile emergency planning zone.

FIGURE 17.1 Probability of Exceeding Whole-Body Doses Given a Core Melt Accident

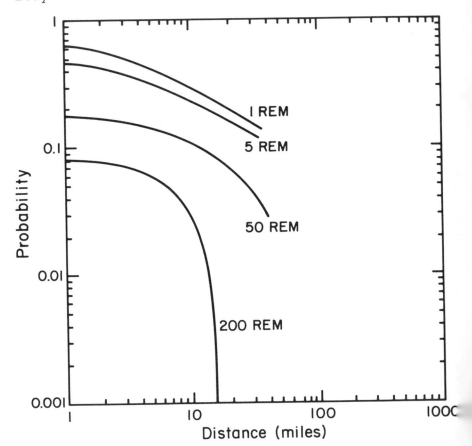

Figure 17.2 is derived from the PRA prepared for the Shoreham Nuclear Power Station and indicates significantly lower conditional probabilities of exposure, particularly at higher doses of radiation.[13] These results are indicative of a recognition that the reactor safety study source terms may be overestimated. The NRC has asked the American Physical Society to reevaluate the source term. Should there be a significant change in the source term magnitude, it is likely that the NRC will reconsider its planning basis.

FIGURE 17.2   Probability of Exceeding Whole-Body Doses Given a Core Melt Release

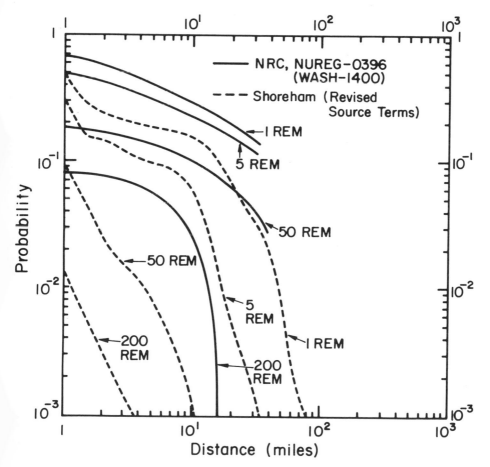

SUMMARY

This chapter only briefly describes the impact of electric power generation on public health. The exact nature of the effects are either imprecisely known, as in the case of air pollution from fossil combustion and related mortality and/or morbidity, or involve complex yet speculative estimates of the probability and magnitude of a core melt accident at a nuclear power plant. Risk comparisons between these two options for future electrical power in New York are complicated by differences in the nature of the risks involved in each stage of the respective fuel cycles.

NOTES

1. Many authors have prepared numerical comparisons of various risks involved in developing and using fossil and nuclear energy. One useful reference is L. D. Hamilton, "The Respective Risks of Different Energy Source," *IAEA-Bulletin* 22, no. 5/6, October 1980, pp. 35–71.

2. "The Direct Use of Coal," Congressional Office of Technology Assessment (Washington, D.C.: Government Printing Office, 1979), 052-003-00664-2.

3. See, for example, "Report on Joint Hearings on the Conduct of the Environmental Protection Agency's Community Health and Environment Surveillance System (CHESS) Studies," U.S. House of Representatives, April 9, 1976 (Washington, D.C.: Government Printing Office, 1976).

4. L. B. Lave and E. P. Seskin, *Air Pollution and Human Health* (Baltimore: Johns Hopkins University Press, 1977).

5. W. W. Holland, A. E. Bennett, et al., "Health Effects of Particulate Pollution: Reappraising the Evidence," *American Jounral of Epidemiology* 110 (1979):525–659.

6. C. M. Shy, "Epidemiologic Evidence and the United States Air Quality Standards," *American Journal of Epidemiology* 110 (1979):661-71.

7. For a full discussion of the risk associated with the nuclear fuel cycle, the reader is referred to the analysis by the National Academy of Sciences, "Risks Associated with Nuclear Power: A Critical Review of the Literature," NAS, April 1979, PB83–158808.

8. "Reactor Safety Study: An Assessment of Accident Risks in U.S. Commercial Nuclear Power Plants," WASH 1400 (NUREG-75-1014), NRC, 1975.

9. Licensee Event Reports are certain reportable occurrences or plant incidents as defined in the Energy Reorganization Act of 1974, Section 208. They are required to be filed by licensees with the NRC.

10. "Using PRA to Assess Nuclear Power Plant Operating Events," Nuclear Safety Analysis Center, August 1982.

11. "The Biological Effects of Ionizing Radiation—BEIR III" (Washington, D.C.: National Academy of Sciences, 1980).

12. "Planning Basis for the Development of State and Local Government Radiological Emergency Response Plans in Support of Light Water Nuclear Power Plants," USNRC, USEPA NUREG-0396, EPA 5 20/1-78-016, 1978.

13. "Draft Report, Ex-Plant Consequence Assessment, Shoreham Nuclear Power Station, PRA," prepared for Long Island Lighting Company by Pickard, Lowe and Garrile, August 2, 1983.

18

Nuclear Safety and Environmental Issues ————————

Alan T. Crane

If nuclear power is to have a future, it will be only if a general consensus develops that it is a safe, reliable, environmentally benign, and economically sound. Twenty, even ten years ago, this consensus existed, but no longer. Virtually every facet of the nuclear fuel cycle is controversial to some degree, and some people reject the entire concept on philosophical grounds. Despite the controversy, however, the early promise has been partially borne out. Some reactors have performed very well, meeting the expectations of their developers and providing considerable benefits to the users of electricity. This is the dichotomy that led to a recent Office of Technology Assessment (OTA) study, *Nuclear Power in an Age of Uncertainty*,[1] from which this chapter is largely drawn.

There is little question that under normal operating conditions, nuclear power affects the environment much less than an equivalent amount of coal mining and combustion. The real questions are what happens under abnormal conditions and how often do these occur. The major areas of controversy are with the safety of the reactors, transportation and disposal of spent fuel, the linkage with proliferation of nuclear weapons, and high-voltage transmission of electricity from remote sites.

The primary focus of this chapter is on reactor safety. The other controversial issues will be considered briefly. It is worthwhile to note that all of these issues involve problems that might occur but have not yet on any significant basis and may never. The accident at Three Mile Island was a major catastrophe for the utility and the nuclear industry, but the effects on the public were very slight. Very little spent fuel will be shipped and disposed of for 15–20 years. No terrorists or nations have diverted nuclear materials from the fuel cycle of power reactors in order to make weapons. The aesthetic impacts of power transmission are

213

real, but research into health effects from proximity to the transmission lines shows negative or ambiguous results. In addition to these secondary nuclear issues, this chapter will briefly discuss some of the impacts of other energy technologies that should be borne in mind if nuclear power is not an option for the future.

The unique safety risks of nuclear reactors stem from the radioactivity in the core, and in particular from the fission product decay that continues after the chain reactor is shut down. As long as the radioactive material is contained in the core, a major release is impossible, but if the decay heat is not removed, the core can overheat and lose its integrity. This possibility has led to an elaborate set of controls and protective systems to shut down the chain reaction, keep the cooling water flowing, supply emergency cooling water if the normal supply is unavailable, and contain the fission products if the cooling fails anyway.

One can postulate a series of events that would incapacitate all these protective systems and allow a major release with devastating consequences. According to an analysis by the Sandia National Laboratory, such an accident could, under certain conditions, lead to 100,000 deaths within a year or, under an alternative scenario, $300 billion in damage.[2] These results may or may not be correct, but clearly any major release is unacceptable. This requirement has led to a continual escalation of safety improvements such as more stringent quality control, greater margins of safety, and further redundancy of systems. These changes have been one of the major causes of the skyrocketing costs of nuclear plants under construction. Thus, safety and costs are closely linked.

Concern over major accidents has also been one of the chief causes of the loss of public support for nuclear power. Three Mile Island had a profound effect on how people view nuclear power. Many in the vicinity of the plant were very worried, and their fear was vividly communicated to the rest of the country because of the drama of the accident. Even though the releases of radioactivity then were small and evidently had negligible consequences, the event drove home the fact that reactors could be a threat. That same event also made it clear that an accident causing extremely expensive damage to the reactor was possible even without harming the public and, in fact, was much more likely. This revelation was quite alarming to utilities, public service commissions, and investors.

The consequences of the worst possible accident are only one part of the overall evaluation, however. If such an accident can be made so fantastically unlikely for each individual reactor that collectively it is very unlikely that the world would experience one in many thousands of years, then that risk can be properly considered acceptable (assuming there are reasonable benefits involved). Murphy's Law (if something can go wrong, it will) has a lot of truth, but it applies only when the time period is long relative to the risk. Since the present reactor technology is transitional and will probably be replaced by something quite different in a few decades, the overall risk from reactors can be made extremely low *if each and every reactor can be held to very high standards*.

Nuclear proponents feel quite strongly that the above analysis accurately represents the level of risk involved. Unfortunately, this contention cannot be proved. It is a judgment based on incomplete data and imperfect analytical techniques. It may be right, but there is still room for doubt. The technology is still quite new. Only about eighty power reactors have been operated in this country, many of them one of a kind. It seems plausible that some of these have undiscovered design errors that might become important under some conditions. The lack of a direct readout on the position of the pressure relief system valves at Three Mile Island is an example (the instrumentation told the operators that the valve was closed when in fact it had stuck open, releasing cooling water). Some plants may have undetected construction errors, for example, poorly poured concrete with too many voids that may leave a structure weaker than its designers had calculated. Inadequate maintenance may render some systems inoperative, such as the circuit breakers in the reactor trip system at Salem. There is still a long list of unresolved safety issues before the Nuclear Regulatory Commission that may eventually indicate that some plants are inadequately protected. Probabilistic risk assessment (PRA), the closest technique we have to quantitative proof, still suffers from an inadequate data base for the failure of equipment and operator error, and is too dependent on the skills and perspective of the analyst. Even the use of actual operating data gives conflicting results. One study of incidents indicated that the risks at existing plants may be significantly higher than the industry has assumed.[3] A critique of this study found that these probability estimates were high by a factor of thirty.[4]

Thus, it has to be admitted that we really do not know what all of the risks are. None of these deficiencies in our knowledge is terribly serious by itself. There are so many levels of safety that if one system is disabled, several others should still be available. Clearly the risk is not high. There are too many reactors around the world that are operating well and occasionally easily riding out the kind of problem that could lead to accidents in improperly designed plants to worry seriously that any individual reactor is a major threat to its neighbors. It is plausible that over the next several decades, a reactor somewhere in the world may suffer an accident resulting in the release of sufficient radiation that a local evacuation should be ordered. A massive release is much less likely, but it cannot be said with certainty now that the risk is so low as to be negligible, or whether it is of a low to moderate level. However, it is significant that those most familiar with the technology are universally convinced that it is more than adequately safe. In the past, this argument lacked credibility because these were the same people with a vested interest in promoting the technology. This criticism is no longer valid. The industry is laying off, not building careers. People with strong reservations would have little to lose by speaking out. Under such conditions, the total lack of defectors is a strong indication that there is nothing important to speak out about.

Over time, these uncertainties in our assessment of the risk will diminish, and the technology will improve. Specific weaknesses will be identified and corrected. Operator training will improve so that plants should run more reliably and safely. The Institute for Nuclear Power Operations (INPO) has been organized to facilitate the exchange of information among plants and raise the standards by which they are operated. Further analytical work may show that some of the assumptions were much too conservative. For instance, the Three Mile Island accident showed that far fewer radioactive isotopes of iodine and cesium were released than had been expected. If the postaccident "source term" is smaller, the worst-case accident should have correspondingly less drastic consequences, and many accidents that seemed to threaten public safety could be shown to have no such consequences at all. One recent study calculated that post accident releases would be less than 10 percent of what had been assumed.[5]

These improvements, however, may be insufficient to restore confidence in the safety of nuclear plants. Unless there is a long period of reasonably uneventful operation, people will note the mishaps at plants anywhere and worry that they may be precursors to an accident that might occur at their local plant. In the past there have been far too many of these incidents. Besides Three Mile Island there were the failures to insert control rods at Salem and Browns Ferry, the fire at Browns Ferry, and the steam generator tube break at Ginna. All these made national headlines. People may be even more concerned over more local but frequent occurrences such as equipment failures that initiate scrams, poor reliability, construction errors, low-level radioactive releases, and the heavy fines levied by the NRC for management inadequacies at several utilities. All of these incidents undermine confidence in the safety of plants. All plants are hostage to the ones with the weakest performance.

If there is one key to restoring the viability of nuclear power, it is in improving industry performance both at existing plants and in the remaining construction programs. Many of the present difficulties stem from the immaturity of the technology and the unreadiness of many utilities and contractors and the NRC to meet the great wave of orders that came in the late 1960s. These mistakes would not be repeated on new orders. The industry has learned a great deal about how to design and construct nuclear power plants. These lessons would be incorporated in any new plants. St. Lucie Unit 2 is probably the best model for how a new generation of plants could be managed. Construction on this plant was begun in 1977 and completed in six years.

St. Lucie and a few other plants show that the technology has largely matured. However, there will not be any new orders unless public confidence is restored, which can only be accomplished by a convincing demonstration that the risks are truly insignificant. This demonstration could start with a full independent audit of each plant's condition. Was the plant, in fact, built in accordance with its blueprints and to its specifications? If important errors are found, obviously they should be fixed. Next would be a complete PRA for the plant. Over twenty plants, including Indian Point, have undertaken these analyses, which are very expensive and laborious. They can, however, be very much in the utility's own interests if they uncover design errors or suggest simple modifications that can reduce downtime or the risk of a damaging incident.

Such steps should eliminate many potential surprises resulting from inadequate hardware. Other steps would be aimed at the human factor. Improvements in operations at some plants still seems painfully slow. Required above all is the recognition by each utility's upper-level management that operating a nuclear plant is a complex and demanding task that depends on a major commitment to excellence. Mediocre people making short-sighted decisions will not produce a good plant. Operators and other plant personnel must be exhaustively trained; such training should include considerable time on a simulator, preferably right at the plant. Maintenance must border on the fanatical to prevent deterioration and forestall downtime later. These practices should result in a safer, more reliable plant and should pay off directly with superior performance. In some cases, achieving these gains will require a considerable shift in thinking from traditional utility management practices. It is encouraging to note that these shifts do appear to be occurring. Recent experiments by public service commissions with rewards for more reliable performance are likely to help.

Equally important, public perception of safety should improve from both the better performance and the increased confidence in the quantitative analysis showing the risk is indeed very low. The role of the critics in communicating this improvement should be noted. Obviously there are many critics who do not want any more nuclear power plants under any conditions, but there are others who have specific concerns with the technology and its management. Special efforts should be made to address these concerns. These critics tend to be the most technically knowledgeable and give the antinuclear movement much of its credibility. As their concerns are addressed, their opposition will diminish. The media might also lose some of their fascination with the sensational aspects of nuclear energy. Once the controversy is lowered from the public's view, the public's apprehensions will lessen.

It is of course possible that these efforts will be inadequate, that performance will not improve markedly, that further analysis will show further problems or that opposition will remain high anyway. If nuclear power appears to be a necessity—perhaps because power demands grow rapidly in the 1990s, or because the environmental effects of coal become too difficult to manage—alternative approaches may have to be considered. Ultrasafe reactors, as discussed by Dr. Weinberg in Chapter 13, might have significant advantages. Since safety is such a concern and has important ramifications for cost escalation, regulatory stability, and public acceptance, a new reactor with safety characteristics that prevent the occurrence of any sequence of events that could lead to a massive release of radioactivity should prove very attractive. There are two cautionary notes, however. First, we do not know for sure if such a reactor will realize its design goals. We have considerable experience with lightwater reactors, but very little with other types of reactors. Closer scrutiny may reveal flaws in the concepts. Second, there is no real evidence yet that an ultrasafe reactor would result in better regulation and increased public acceptance. No reactor can be perfectly safe, and even though these reactors should in principle be credited for their inherent safety, practice might prove different.

Even if it is conceded that the risks reactors pose are acceptable, there will still be much opposition to nuclear power, especially as long as the waste issue remains unresolved. A program has been put into place by the Department of Energy as a result of which a repository should be ready to accept waste by about the end of the century. The program is ready to characterize several potential sites in detail, but intense local opposition to each may be a significant problem. The program has provisions for compensating residents, and if the matter is approached wisely so that no one area feels it is being unfairly singled out as the dump for the rest of the nation, agreements may be reached. The technical problems are also considerable, but there do not appear to be any that cannot be resolved to the extent necessary to meet EPA standards, assuming a suitable site is found. Once the waste is in the ground, the worst conceivable accident appears to be a slow leakage and a relatively modest addition (much less than 1 percent) to other sources of radioactivity.[6]

Opposition to high-voltage power lines can be nearly as intense as that to nuclear waste depositories, but the reasons appear to be slightly different. A waste disposal site would be to all appearances just another industrial or research facility. It is the potential risk to health if the material proves not to be contained adequately that is of greatest concern. In the case of electric transmission, the evidence of any health effects is ambiguous at best even for the new 765-kilovolt (kV) lines.[7] Clearly, people feel threatened by the lines, but the health effects may be a surrogate for all the other reasons as well: the aesthetic intrusion into the landscape, interference with television reception, and a sense of outrage over one's land being taken, seemingly arbitrarily.[8] It is likely that more power lines will be needed, especially if purchases from Canada are increased. The higher the voltage, the fewer the lines needed to carry the same power, but the greater the impact of each individual line. Obviously, there is no easy answer to the problem, nor any way to keep everyone happy, but it does seem incumbent on utilities and the state to search for better mechanisms for locating lines and securing rights to the land.

The problem of proliferation is harder to define. No one expects a U.S. utility to make a nuclear weapon from its own spent fuel, and as long as there is no reprocessing there are not credible means by which terrorists could do so. Since the infrastructure for reprocessing in this country is at least several decades away, that should not concern us greatly. The real concern with the use of nuclear energy by the U.S. is a moral one. It is possible for another country to divert material from its own commercial program to make a bomb. Under most conditions this would not be the preferred route. For instance, a small plutonium production reactor can be built quite cheaply and secretively, and produce a better grade of material. However, having a power reactor and the expertise to operate it could make it easier, especially for a developing country, to prepare itself for a weapons capability should it later desire to produce it.[9] A worldwide renunciation of nuclear energy would certainly reduce this possibility, but the usefulness of a unilateral U.S. renunciation is highly questionable. In fact, it is at least as likely that it would be counterproductive, driving developing countries to less principled suppliers. The industry itself has strong incentives for preventing any exercise of the potential linkage between commercial power and weapons. Such proliferation would be a devastating blow to the industry, and any means to minimize this possibility should be given careful consideration.

The nuclear industry has been beset by many problems. Some might be called accidents of history; for example, the sudden drop in electric demand growth that made so many of the plants that had been ordered unnecessary, and the high inflation and interest rates that made the rest so expensive. Others have resulted from inept regulation and unscrupulous opponents, but these have not caused the problems as much as made real problems worse. It is clear by now that the industry itself is the source of many of its own problems: the immaturity of the technology, the overconfidence that led to growth that started too soon and happened too fast, the underestimation of the complexity of the task, the arrogance that any problem could be solved and had to be for the atom to achieve its rightful place.

In summary, too many reactors are operating reliably, economically, and safely to believe that the technology is hopeless and that the nuclear option has been written off. The industry is shrinking and changing but it won't disappear. It has already overcome many of its past problems and the next generation of reactors should be significantly improved. Considering the industry's great will to live and belief in itself, it is possible to think of a resurgence, not in the next few years but by the end of the century.

If, however, nuclear power is not an option for the future, utilities will be looking for alternative types of generating facilities. Very few observers of our energy system think that the nation's demand for electricity will not grow, except perhaps under implausibly intrusive government policy or unexpectedly high prices. New capacity will be needed both to meet load growth and to replace worn-out generating capacity. If this capacity is not to be met by nuclear-powered plants, the most likely choice is coal. As discussed below and in Chapters 16 and 17, coal has very serious consequences of its own. A brief description will help place those of nuclear power in context, but this is not intended to be a comparison. In fact, environmental comparisons between coal and nuclear tend to be somewhat sterile. The kinds of impacts, risks, uncertainties, and opinions each involves are too different to permit anything but a vague and qualitative comparison.

It should be noted first that the coal plants built now or in the future will be much cleaner than those of the past. Belching smokestacks and sulfurous fumes no longer exist. All the known emissions of a coal plant that are present in sufficient quantities to cause human health problems are controllable by various means.

What remains controversial are the effects of acid rain, chronic human health effects owing to low levels of pollutants, and climatic changes as a result of carbon dioxide. Briefly, with regard to acid rain, substances such as sulfur dioxide and nitrogen oxide that do come out of the stacks are transformed in the air into sulfuric acid or nitric acid among other things. This transformation is exacerbated by tall stacks that can push the pollutants high into the atmosphere. There they can remain suspended for long periods of time, allowing a higher fraction to be converted to acids and also allowing long-distance transport. Neither transport nor transformation is yet well understood, which makes the determination of the source of the acid rain in any specific area very difficult and uncertain. Since all new plants have strict controls on the amount of sulfur they can emit, the bulk of the emissions is from older plants, whose restrictions are lenient or, in some cases, lacking altogether. As concern over acid rain has mounted, attention has focused on reducing emissions from these older plants. Retrofitting an old plant with a modern fuel-gas desulfurization is very expensive, however. This results in some very difficult decisions: we are reasonably sure that acid deposition is causing significant damage, but because the damage may be several states away from a source that cannot be identified precisely, the polluter would derive no direct benefits from the large investment needed to clean up the plant. Several proposals are being considered by Congress that would try to strike a compromise.[10]

Public health effects are harder to quantify than are the effects of acid rain. The present ambient air quality standards were set on the basis of clinical tests of damage following exposure to measured levels of various pollutants. Since few people are routinely exposed to levels higher than these levels, there should be no impact on human health. However, epidemiological studies using complex statistical techniques have linked current levels of sulfate concentration to tens of thousands of premature deaths annually in this country. These studies are quite controversial and contain many uncertainties. However, the results are not implausible because people already sick are susceptible to any further illnesses, and if they die sooner than expected, it would be of bronchitis or asthma or another "normal" cause that would not be ascribed to pollution. Sulfates are the same form of sulfur associated with acid rain. Although they may be just a surrogate for all the pollutants that can cause damage, it seems reasonable to assume that they would be one of the main actors if these analyses are correct. Hence the same measures taken to reduce acid rain should also reduce the human health effects.[11]

The final pollutant of concern is carbon dioxide. This is an inherent product of any form of combustion and is not in itself chemically damaging. However, it can produce what is known as the "greenhouse effect" in the atmosphere. Carbon dioxide is transparent to the light from the sun but reflects back the longer infrared wavelength being reradiated from the earth's surface as does the glass in a greenhouse. Trapping this heat will gradually warm the entire world. Eventually, this effect could become extremely serious. Drastically changed weather patterns would profoundly disrupt agriculture even if the shifts ultimately proved beneficial. Some populated areas might become uninhabitable. Enough of the polar ice caps could melt to raise the sea level significantly, flooding coastal areas. None of these major changes would be much in evidence for several decades or more. The reason for worrying about them now is that no real cure is evident; the cause is worldwide, which will necessitate coordinated international action even to stem the problem; and the changes required will take decades to implement. Although an immediate call to action is not warranted because our understanding is still inadequate, it should be noted that in the coming decades a drastic reduction in the use of fossil fuels may be required. The more nonfossilized energy sources are emphasized now, the slower will be the buildup of carbon dioxide, and the easier will be the transition. This is probably the most compelling argument for maintaining the nuclear option, at least in the long run.[12]

Thus coal, despite its abundance and the enormous improvements in technology over the past few years, is not a particularly attractive alternative to nuclear. It is probably a major cause of acid precipitation, it may well be the cause of a staggering human health problem, and it appears to be leading to potentially monumental environmental changes.

This analysis does not, however, inexorably commit the country to a nuclear future. Although U.S. oil production is almost certainly in a permanent decline, more deposits exist both at home and abroad. Natural gas production may even increase for several years, possibly longer. The favorite alternative options of nuclear critics, however, are the use of solar energy and conservation. The latter is the one true success story in energy since the oil embargo of 1973–74. It is quite remarkable how many ways people found to use energy more efficiently when faced with vastly higher prices, and many further opportunities still exist. Even some forms of conservation, however, have negative environmental impacts. As houses have been tightened up to reduce heat loss, ventilation also diminishes. Many houses have been made so tight that indoor pollutants accumulate to dangerous levels. Various toxic chemicals such as formaldehyde are given off by materials in the house. Radon is released from the ground or concrete. Combustion products are produced by cooking or any form of heating except electric or solar. Since most people spend most of their time indoors (especially those most vulnerable to pollution—the very young, the sick, and the aged), indoor pollutants can present a far more serious problem than ambient pollution outside. Ventilating heat exchangers are available, but they are quite expensive and waste a significant amount of energy. All of these problems can be solved; rather than take complex precautions, however, most people are likely to prefer a slightly leaky house that uses somewhat more energy.[13]

Of the many forms of solar energy, only hydro-power and wood are now in widespread use. Wind power is growing rapidly, but this is likely to start to level off in a few more years because of siting constraints. Hydro-power also has limited additional potential. The other alternatives are still too expensive, although the potential for some—photovoltaics, for example—is great for the twenty-first century.[14] Many people use firewood quite happily though there are indications the use has leveled off. Commercial and industrial use will still grow. Unfortunately, wood is also by far the most dangerous energy source we have. Chain saws, falling trees, sprung branches, transportation of heavy loads, and splitting wood cause many injuries and deaths each year. Smoke is a major problem in some areas, and wood stoves can be dangerous in tight homes.

The purpose of this chapter has not been to show that nuclear is the best or even one of the better energy sources. There is no question that nuclear power presents risks and problems to society. Although the risks and problems are often exaggerated, sometimes dramatically, they are real. Doing without nuclear, however, only changes the overall risk to society, and perhaps not for the better. With this in mind, maybe our society can come to a rational decision on what if anything we want from nuclear energy.

NOTES

1. Office of Technology Assessment, *Nuclear Power in an Age of Uncertainty*, OTA-E-216 (Washington, D.C.: U.S. Government Printing Office, February 1984).

2. Sandia National Laboratory, *Technical Guidance for Siting Criteria Development*, U.S. Nuclear Regulatory Commission NUREG/CR-2239 (Washington, D.C.: U.S. Government Printing Office, December 1982).

3. M. W. Minarick and C. A. Kukielka, *Precursors to Potential Severe Core Damage Accidents: 1969-1979* (Oak Ridge, TN: Science Applications, 1982).

4. Institute for Nuclear Power Operations, *Review of NRC Report: "Precursors to Potential Severe Core Damage Accidents: 1969-1979"* (Atlanta, GA: Institute for Nuclear Power, 1982).

5. American Nuclear Society, *Summary Report Special Committee on Source Terms* (Chicago, Ill.: American Nuclear Society, September 1984).

6. Environmental Protection Agency, *Draft Environmental Impact Statement for 40 CFR191: Environmental Standards for Management and Disposal of Spent Nuclear Fuel, High-Level and Transuranic Radioactive Wastes*, EPA 520/1 82 025, December 1982. See also, Office of Technology Assessment, *Managing the Nation's Commerical High-Level Radioactive Waste*, OTA-0-171 (Washington, D.C.: U.S. Government Printing Office, March 1985).

7. Electric Power Research Institute, *Electromagnetic Fields and Human Health*, EPRI *Journal* 9, no. 6 (July/August 1984).

8. Louise B. Young, *Power over People* (New York: Oxford University Press, 1973).

9. Office of Technology Assessment, *Nuclear Proliferation and Safeguards* (New York: Praeger Publishers, 1977).

10. Office of Technology Assessment, *Acid Rain and Transported Pollutants*, OTA-0-204 (Washington, D.C.: U.S. Government Printing Office, June 1984).

11. Office of Technology Assessment, *The Direct Use of Coal*, OTA-E-86 (Washington, D.C.: U.S. Government Printing Office, April 1979).

12. Massachusetts Institute of Technology Energy Laboratory, *Global Energy Futures and CO_2-Induced Climate Change*, MIT-EL 83-015 (Cambridge, MA: MIT Energy Laboratory, November 1983).

13. Office of Technology Assessment, *Residential Energy Conservation*, OTA-E-92 (Washington, D.C.: U.S. Government Printing Office, July 1979).

14. Office of Technology Assessment, *New Electric Power Technologies: Problems and Prospects for the Nineties* (Washington, D.C.: U.S. Government Printing Office, July, 1985).

PART V

PUBLIC PERCEPTIONS OF THE ELECTRIC UTILITY INDUSTRY ————————

Public perceptions and attitudes toward the electric utility industry are not usually included in discussions about electrical energy. Yet in recent years these perceptions and attitudes have become even more important to the future of the industry than its engineering, economic, or institutional factors. For it is precisely these perceptions and attitudes on the part of the public that may have the greatest impact on the political and regulatory framework in which the industry operates. Scientists, engineers, economists, lawyers, and other professionals may be reluctant to admit that public perceptions and attitudes, whether basically right or wrong, could dominate any particular technical issues. However, those familiar with the nuclear power question are aware of the influence that generally negative public perceptions and attitudes have had on recent developments in that industry. Although much of what has been written concerning public attitudes about electrical energy has focused on the nuclear issue, this is only one of many important issues related to the generation and use of electrical energy.

The three chapters in this section focus on different aspects of public perceptions toward the utility industry. Donald DeLuca provides a general overview of some of the important concepts of survey research and discusses what is required in order to make valid generalizations to the whole population from sample surveys. Such surveys are important because they are probably the most common method for measuring public attitudes and perceptions. In addition to statistical concepts, he raises issues that surround wording, context, values, salience, conflict, and consensus, illustrating his discussion with examples from energy-related surveys. DeLuca also provides a useful historical review of public perceptions and attitudes towards nuclear energy.

225

As a follow-up to Deluca's methodological inquiry, Steve Barnett summarizes the results of in-depth group interviews with sixty-four consumers throughout New York State and provides some specific examples of current attitudes among the public. According to Barnett, consumers of electrical energy in New York State are a very dissatisfied lot in terms of their attitudes toward energy conservation, the electric utilities, and the state government's role in regulating utilities. Barnett offers some specific recommendations to help utilities and government officials change these negative consumer attitudes.

The main purpose for gathering such survey data and in-depth interviews is to learn more about the public's perceptions and attitudes, though such knowledge is especially useful if it leads to changes for the better. Such surveys may also be used to predict the likely outcomes of alternative policies by the government or utilities.

In the final chapter in this section, Howard Axelrod outlines an approach to predicting regulatory crises that he believes could be beneficial to the managers of utilities. The five basic regulatory concepts that he deals with are equity, risk, incentives, economic impact, and the doctrine of "used and useful." At this time, there are no reports of any tests of his model.

19

Survey Research on
Energy-Related Issues ——————
Donald R. DeLuca

Survey research is a powerful tool for providing precise, quantifiable, and generalizable information about public sentiment, preferences, motivations, attitudes, and behaviors on many topics. Understanding public opinion on broad, complex, and relatively nonsalient policy issues such as the nation's energy situation, the environment, and technology requires sophistication on the part of the survey organization that designs the study and collects the data, the analyst who interprets the findings, and decision makers, as well as other members of the thinking public who receive the information. Some of the elements necessary to obtain such sophistication are discussed below in order to help stimulate more informed poll watching.

Last summer, a disarmament conference was held in Italy to exhort the superpowers toward greater arms control. The Soviet delegate was challenged to explain why peace committees in the Soviet Union, unlike their Western counterparts, never criticize their own government's policies and missile installations pointed at Western Europe. His answer was: "Public opinion and official opinion are the same thing in our society. They are always the same. We have ways of establishing this link."[1]

It may be surmised, with some confidence, that he was not talking about public opinion polls as a way of establishing that link. In American society, however, there are public opinion polls, a free press, and institutionalized mechanisms for public disagreement and protest, with constitutional guarantees establishing these as a right. Yet there is good journalism and bad, sometimes violent protest, and good and bad polls. Americans' feelings about polls were perhaps best summarized by Phillips B. Ruopp of the Kettering Foundation:

Too often public opinion polls are treated as ends in themselves—a form of gossip half-accepted as gospel—rather than useful tools for probing the complexity of public images, attitudes, and behavior. People love polls; everyone can quote the latest poll relating to his or her particular interest—and most of the time the polls people cite support their own preconceived opinions. And because people love polls, the media love them too. While everyone has access to many polls, too few people seem to know what to do with them.[2]

THE SURVEY RESEARCH DESIGN

The social survey is an elegant research design. It is composed of a rigorous protocol based on scientific probability sampling that yields numerical data at the individual level. If the sample design is based on these scientific principles—and not all so-called surveys are—results are generalizable to the entire population from which the sample was selected. Hence, it is possible to interview as few as 1,500 adults and have the results very accurately reflect the attitudes, motivations, preferences, or behavior of the entire adult population in the country. If well designed, a survey can provide more accurate data than can a census of the entire population. Statistical theory provides an exact estimate, at a given level of precision, of how well the sampled finding represents the population as a whole: the standard error of the estimate. It is this characteristic of the survey research design that makes it such a powerful tool.

In some ways, the survey research design is the only methodological technique available for uncovering such information. There are other research designs that contribute to understanding phenomena at the individual level of analysis. However, the survey research design provides at least one thing that these other research designs do not: quantifiable data that are generalizable. Focus groups, planning cells, quasi-experimental laboratory designs, or case studies may yield more in-depth information, but results from such research suffer from a lack of generalizability.

However, there are a number of problems associated with the survey research design that greatly affect the interpretation of survey findings and use of survey data for decision making. These problems derive from limitations of theory and measurement inherent in the technique itself. Also important, and discussed later in this chapter, is the need to anchor firmly conclusions drawn from survey research to the broader context of the salience of the issue examined, personal experiences, and the basic human values of the respondents.

The potential *total* error from gathering survey data goes well beyond that which may be accounted for by sample design.[3] Other potential problems that require careful attention are measurement error (e.g., instrumentation bias, issues of validity and reliability, response bias); implementation error (e.g., response rates, interviewer error, selection bias); and data-processing error (e.g., editing, coding, scaling, recoding). Yet when news commentators or newspapers report on a poll, they commonly conclude by stating that the findings of the survey are , e.g., "accurate to within plus or minus three percentage points." This refers only to the more scientific part of survey research, which lends itself most easily to quantification and statistical manipulation. Although the survey research design allows precise, quantifiable, and generalizable responses based on the questions asked and the topics studied, the art of asking questions, especially on complex topics, with low immediate salience to the respondents answering the questions, requires experience, creativity, and sophistication.

It is perhaps obvious that emotionally laden words affect questionnaire responses; however, often even very subtle wording changes in questions will greatly affect the pattern of responses.[4] Whether the questions asked are filtered or unfiltered (for example, whether the respondent is first asked if he or she has enough information to have an opinion, or, even if he or she does have an opinion, before being asked what that opinion is) will yield a quite different set of responses. These and many other methodological considerations are very important for understanding the results of survey research, especially when confronted with a headline poll presenting a single finding.

Survey questions appear in many forms, wordings, and contexts.[5] Frequently, when one survey organization asks a question on an energy topic (e.g., nuclear electric power), it will be examining only one aspect of the nuclear electric power debate. Another survey organization may be asking questions about yet a different aspect of that debate on the very same day. Under these circumstances, it is not surprising that the two surveys, conducted at the very same time, contain different percentages of the population that either favor or oppose future development of nuclear electric power. Without having access to the many survey findings through time on questions related to nuclear electric power, one is perhaps tricked into relying too heavily on that single finding for information about public sentiment, thus becoming the victim of the headline poll. As Albert Cantril states, reporting a single percentage to summarize the complexity of public opinion on an issue is clearly inadequate:

> The public opinion poll has the capacity for the precise measurement of public sentiment, but it is a tool that we use bluntly all too often. For

reasons of convention within the profession as well as pressures operating on it from outside, pollsters persist in a preoccupation with producing that single number—percentage of the public—that must somehow bear the burden of conveying the state of opinion on an issue, no matter how complex it is or how riddled with contradictions public opinion may be.[6]

PERSONAL EXPERIENCE, VALUES, AND SALIENCE

To understand public opinion about energy, especially when examining large-scale national or regional surveys, several contextual factors should be kept in mind. Surveys measure and aggregate the variegated and segmented opinions of the polity. The polity is very diverse. It includes people from all walks of life.

Each individual is in a particular social milieu and thus each person's experiences are very different. When respondents were asked about their attitude toward the energy crisis in the mid-1970s, it mattered a great deal whether they came from an area where schools and factories were closed; gasoline lines were long; utility bills had a sudden and rapid increase in price; and what, where, and how information about the energy crisis was received. Individuals also differ in their personal responses to these experiences. Two individuals with the same life circumstances may approach a given problem quite differently and, therefore, may have very different attitudes on the same topic. For example, someone with altruistic motives may have viewed the energy crisis as an exciting challenge, and another individual may have viewed it as such a serious threat that the reality of its very existence was denied.

Survey findings will be better understood and used more accurately when the analyst is aware of these complexities. Maintaining common sense and avoiding truisms about the public also enhance the accuracy of one's perception of public sentiment about the nation's energy situation.

Another contextual factor that is important for understanding attitudes toward energy-related issues is the salience of energy matters to the respondents of the survey. Gallup polls have been asking a standard, open-ended question for over four decades: "What is the most important problem facing the country today?" Only once in forty years, during the peak of the oil embargo crisis in the mid-1970s, did energy emerge as the nation's most important problem. Rarely did energy even appear anywhere on the long list of problems volunteered by respondents. Energy matters, compared to the economy, defense and national security, or employment (or even compared to daily living conditions and personal family matters), are not very salient issues for most people.

The importance of public opinion for policy varies by the kind and level of policy decision being considered. It makes a great deal of difference, for example, whether the attitudes being examined are about nuclear electric power or about some energy conservation matter. Expression of a generalized attitude toward nuclear electric power does not directly relate to an exact behavior associated with it: nothing can be done immediately. Whether or not to insulate an attic is something that requires direct personal involvement and thus has high salience to the individual.

The responses given to questions about attitudes on surveys are also conditioned by the human values of the people interviewed. Values have a technical definition that psychologists use, as a term referring to underlying orientation. One forms values through various cultural processes. They are not easily modified. In fact, it is ethically questionable whether there should be any attempt to modify basic values. One's values shape those aspects of life that relate to goals that individuals consciously or unconsciously perceive to be important (for example, a comfortable lifestyle, family security, national security, a world of beauty or equality). Values also help shape the actions people feel they should take to achieve each of these goals. The values people hold, as well as behaviors associated with living by them, perhaps obviously condition people's responses to survey questions.

Risk assessment involves calculating not only the probability of the occurrence of an event but also the consequences of it: the probability multiplied by the magnitude of the consequence. One group, by its values, may emphasize the low probability of an event's occurrence and base their attitude on that. Others may emphasize the high consequence at any non-zero probability. In fact, that is precisely the situation found when attitudes toward nuclear electric power are examined: significant numbers of people in the population view the issues associated with nuclear power quite differently.

Survey data will provide more useful information if all of the above-mentioned factors are taken into account: understanding the power and limitations of the survey research design; using caution in inferring generalizations based on a single headline poll; and recognizing that contextual differences in personal experiences, salience, and basic human values add complexity to survey results. Many of the problems ascribed to survey data are attributable to expectations about the information they can provide. Polls are not a public referendum. Despite these limitations, attitudes are related to action: attitudes do not, however, fully explain behaviors. Measures of attitudes from surveys report what people say, not necessarily what they do. The more abstract the situation, the less clear the link between attitudes and behavior.

CONSENSUS, CONFLICT, AND TRENDS

Before turning to an analysis of the data, it is important to emphasize that attitudes of Americans on energy-related issues are complex and sometimes inconsistent. Part of the explanation for this is the fact that the energy issues themselves are complex, often technical, sometimes controversial, and unlike some issues, involve a curious mixture of public and private decision making. Also, the above-mentioned factors of salience, personal experience, and human values further confound attempts to simplify descriptions of energy attitudes.

Elaborating on a schema described by Daniel Yankelovich, patterns of energy attitudes, in the aggregate, may be placed into two broad categories: consensus and conflict. A *consensus* of opinion on energy issues may be designated as aggregate attitudes that show a pattern of widespread agreement; responses do not vary much with variations of question wording; and only minimal differences in degree, not direction, between various demographic subgroups are evident. The opinions are consistently and firmly held. A *conflict* of opinion exists when the aggregate patterns of attitude measures show sharp cleavages among the entire sample or between demographic subgroups; the responses reveal serious misinformation; and virtually the same or similarly worded questions produce different answers when the phrasing of the questions is slightly changed.[7]

Energy attitudes have most often displayed a pattern of conflict: attitudes marked by disagreement or dissent, where opinions appear confused, inconsistent, based on misinformation. Two types of conflict have been elaborated, internal and external. *Internal* conflict refers to individual respondents who are really not sure how they feel about a given issue. The respondent may have an opinion today, which may be slightly different tomorrow, then very different the next day—the opinion may change easily. *External* conflict exists when individual opinions are firmly held, but over the aggregate of the population, there are polarized groups who have strong, opposing opinions. Attitudes toward nuclear electric power are an example.

In order to help unravel the complexity of attitudes toward energy issues and to avoid the pitfalls of drawing unwarranted conclusions based on a single headline poll, it is wise to examine trends in attitudes over time when such data are available. It is possible to analyze trends in data over a substantial period of time for attitudes toward nuclear electric power: questions have been asked on this topic regularly since the 1950s.

Even with such data in hand, however, it is sometimes difficult to express the complexity of the issue in a single headline story. The following analysis by Robert Mitchell shows the difficulty of capturing, in a headline, the essence of public sentiment toward nuclear electric power shortly after the accident at Three Mile Island:

> The interpreter of these trend results faces the classic is-the-glass-half-full-or-half-empty dilemma. Does one emphasize the sharp rise in opposition or the fact that the bottom did not fall out of nuclear's support? Here are the relevant headlines from the pollsters' press releases showing how they handled this dilemma:
>
> Harris release April 26, 1979:
> AMERICANS OPPOSE PERMANENTLY CLOSING ALL NUCLEAR POWER PLANTS DESPITE THREE MILE ISLAND ACCIDENT
>
> Harris release May 3, 1979:
> AMERICANS UNWILLING TO DECLARE A MORATORIUM ON NUCLEAR POWER, DESPITE INCREASING WORRIES ABOUT ITS SAFETY
>
> Gallup Index April, 1979:
> AMERICANS FAVOR "GO-SLOW" APPROACH BUT CONTINUE TO FAVOR NUCLEAR POWER
>
> CBS/New York *Times* release April 9, 1979:
> ONLY SLIGHT MARGIN FOR APPROVAL NOW RUNNING FOR BUILDING MORE NUCLEAR POWER PLANTS...NUCLEAR POWER CLEAR CHOICE OVER HIGH PRICED OIL (summary statements on cover of release)
>
> New York *Times* story April 10, 1979:
> POLL SHOWS SHARP RISE SINCE '77 IN OPPOSITION TO NUCLEAR PLANTS
>
> Each of these headlines summarizes the overall findings of the poll which was being reported, of course, but the trend data were (or should have been) a centerpiece of the story.[8]

Another way to avoid the problems of a single headline poll is to examine in-depth multiple questions on selected issues and seek patterns to explain consensus or conflict in opinion.

ATTITUDES ON ENERGY ISSUES

Social surveys are a research technique that facilitate the study of public attitudes toward, belief in, and reactions to the energy crisis that began with the OPEC embargo in November 1973. There is some public awareness that no energy source is infinite, but experience with actual shortages is relatively new: the post-1973 oil embargo period, characterized by national shortages of gasoline and regional shortages of home fuel oil; the winters of 1976 and 1977, characterized by regional shortages in natural gas used by commercial establishments and for home heating; and the summer of 1979, characterized by a sharp rise in the price of gasoline and fuel oil, followed by reports of record high oil company profits and renewed gasoline shortages. Reactions to and belief in the seriousness of the energy crisis were largely conditioned by these experiences.

Numerous surveys attempted to gauge the perceived seriousness of the energy situation during the 1970s. However, special attention must be paid to two factors: question wording and question format. Question wording alone can explain differences among otherwise identical surveys sampling opinions concerning the energy "situation," "problem," "crisis," or "shortage." This complicates the interpretation of some of the results presented below. Question format problems arise from the difference between questions provided with a full range of response options and those that are open-ended. The following data on perceived seriousness all come from survey questions provided with categorical responses.

After the initial shock of the 1973 oil embargo, which caused major disruptions in the availability of petroleum products, public perception of the seriousness of the energy situation dropped in 1974, followed by a steady, but gradual increase through 1975. Comparable data on this issue come from the American Institute of Public Opinion (Gallup). The Gallup surveys (Table 19.1) show a mild drop in the proportion of the population stating that the situation was serious in June 1979, perhaps reflecting the short period of calm that occurred just prior to the return of gasoline lines (and odd/even rationing in some states) a few weeks later. Demographic breakdowns show few differences attributable to any variable except education: college-educated people were more likely to view the energy situation as very serious. Republicans were somewhat more likely to consider the energy situation as at least fairly serious.

Over half of the sample did not feel that energy was a *very* serious problem, even after the president appeared on television twice in one week claiming that the challenge of solving the energy problem would "require a moral commitment equivalent to war." Of many possible explanations for this finding, only some may be examined through survey research: lack of confidence in the president; political party loyalty; lack of trust in government, business, or the scientific community; ignorance of current events; or blind faith in technology.

One potential explanation for this skepticism was explored in the Roper Organization surveys: the extent to which people believed that the

TABLE 19.1 Seriousness of the Energy Situation

How serious would you say the energy situation is in the U.S.?
(very serious, fairly serious, or not at all serious)

	Very Serious	Fairly Serious	Not at All Serious	No Opinion
1979				
August[a]	47	35	16	2
June	37	36	24	3
February	43	42	13	2
1978				
March–April	41	39	15	5
November	40	42	14	4
September–October	40	40	16	4
June	40	42	13	5
April–May	44	41	12	3
April (early)	41	39	16	4
Group Composition, August 1979				
College graduate	60	31	8	1
High school graduate	41	38	19	2
Grade school graduate	43	35	18	4
Easterner	45	38	16	1
Midwesterner	42	38	19	1
Southerner	51	31	16	2
Westerner	52	35	10	3
Republican	49	41	9	1
Democrat	50	33	15	2
Independent	43	37	19	1

Source: Surveys by the Gallup poll, the latest taken in August 1979, via personal interview.

[a]See Source.

Note: Data are percent responding.

energy shortage was contrived. Table 19.2 shows that, of the four options presented, most respondents chose the one stating "there never was any real shortage." This contrasts sharply with the second most frequent choice—that the shortage was not only real but would get worse in the future. Fewer people saw the shortage as already solved or likely to be solved in the near future.

In order to understand whether the public will accept proposed solutions to the energy problem, it is also useful to determine public perception of the problem's cause. If foreign countries or corporations, rather than American institutions, are held responsible, then potential solutions

TABLE 19.2 Energy Shortage: Real or Contrived?

Here is a list of statements about the gasoline and oil shortage (respondent given card). Which one of those statements comes closest to expressing your opinion?
  A. *There is a very real shortage and the problem will get worse during the next 5 to 10 years.*
  B. *There is a real oil shortage but it will be solved in the next year or two.*
  C. *There was a short-term problem, but it has been largely solved and there is no real problem any longer.*
  D. *There never was any real oil shortage—it was contrived for economic and political reasons.*

	Statements				Don't Know
	A	B	C	D	
1974					
May	21	12	8	53	6
1976					
May 8–15	25	9	9	48	9
November 6–11	26	11	8	46	9
1977					
April 30–May 7	40	15	6	33	6
October 29–November 5	33	14	9	39	5
1978					
April 22–May 3	32	8	9	45	6
October 28–November 4	31	10	9	43	7
1979					
January 6–20	29	8	8	48	7
April 28–May 5	25	12	6	51	6
September 22–29	28	9	7	51	5

Source: Surveys by the Roper Organization on the dates indicated via personal interview.
Note: Data are percent responding.

that the public would find most reasonable require dealing with sovereign governments outside the direct control of the U.S. If individual actors are held responsible, then individuals may believe that they can change the situation by taking personal action.

Of course, question wording will again significantly affect responses. Asking simply, "Who is to blame?" is different from instructing the respondent to assign a level of blame to a predesignated list of response categories; in turn, results will be further affected by the particular list of response options provided. Close examination of the results of many surveys reveals that while sample proportions vary a great deal depending on question wording, rank order or attribution of responsibility does not change much. The data presented in Table 19.3 are directly comparable through time, since these surveys asked the same question on each of five occasions.

Oil companies consistently received a major share of the blame through time. This finding is also supported by other ways of asking the question, including "cue-free," open-ended questions. The sample proportion who blamed oil companies remained stable at around 55 percent during 1974–77, but by 1979 some 72 percent attributed major responsibility to oil companies. Arab countries ranked second throughout the period (except for the initial survey in 1974), followed by the administration, Congress, and electric power companies. The proportion blaming themselves (American consumers) varied through time, although the rank for this response was fifth or sixth every year except 1977, when American consumers virtually tied for second with Arab countries and electric power companies.[9]

Public expectations about the energy problem are particularly important for natural resource and energy planning, because they generally re-

TABLE 19.3   Who Is to Blame for the Energy Problem?

Here is a list of groups who have been mentioned in one way or another as being to blame for the current energy crisis here in the United States. (Respondent handed card.) Would you go down that list and for each one tell me whether you think they deserve major blame for the energy crisis, some blame, or no blame at all?

	June 1979	June 1977	January 1976	February 1975	January 1974
Oil companies	72	55	57	57	56
Arab countries	51	32	37	38	22
Administration	36	24	28	28	39
Congress	34	28	26	26	26
Electric utilities	25	31	29	26	15
American consumers	23	31	18	20	18
Israel	17	10	13	10	11
Environmentalists	15	13	9	11	10

Source: Surveys by the Roper Organization on the dates indicated via personal interviews.

Note: Data are percent accorded "major blame."

flect both previous experience and assessment of current and future situations. Social surveys on expectations may provide insights for instituting effective programs to mitigate problems.

In April of 1979 a Harris poll asked how serious the energy shortage would be in ten years. Of the responses, 55 percent said that it would be very serious; 22 percent said that it would be somewhat serious. This is approximately the same proportion that stated that the energy problem was very serious at the height of the embargo period. The same Harris poll showed that the public expected that coal would replace oil in providing most of the country's electrical needs in the early 1980s; nuclear and solar energy would jointly provide primary electricity needs by the end of the century; and solar energy would emerge as the primary source of electricity needs by the year 2000, although a substantial proportion of the public felt that nuclear energy would still provide the major source of electrical energy by the end of the century.[10]

Many people seem to be aware of the uncertain availability of oil in the future. Since they expect alternative energy sources to meet the demand in both the short and long run, they are relatively unconcerned about the scarcity of oil and gas that is causing "temporary" disruptions in their lifestyles. A Harris poll in April 1979 asked:

> Generally speaking, how optimistic are you about the ability of this country to solve the energy problem by technological discoveries and developments—very optimistic, somewhat optimistic, or not optimistic at all?

Some 86 percent responded that they were very optimistic or somewhat optimistic. Only 7 percent were not optimistic at all, and another 7 percent were not sure.[11]

Since the 1973 embargo and subsequent government policies encouraging (or forcing) conservation, many surveys have been conducted on "public attitudes toward energy conservation." Since the surveys were uncoordinated, the literature is flooded with myriad findings, many of which are contradictory. Without a close analysis of the research design and without having a copy of the data, it is virtually impossible to determine whether disparate results represent valid changes in attitudes or whether they result from a lack of methodological rigor.

Conservation is a complex phenomenon. Since the findings in survey research on conservation attitudes and behavior do not constitute a consistent pattern, only general statements can be made, with no real confidence in the stability, reliability, or even validity of these findings. People do state a preference for regulations and rationing rather than market-pricing mechanisms to achieve energy conservation. Although some surveys show that people profess a reluctant, though favorable attitude toward increasing the austerity of their lives if necessary, most reported behavioral patterns do not support this. A substantial proportion of the public believes that conservation will limit economic growth and depress the current standard of living.

Perhaps the main reason for inconsistent results from surveys on energy conservation is that, as a variable, conservation is quite multidimensional; it involves many separate actions resulting from numerous decisions by many people. Some survey researchers have sought information on voluntary actions, others have measured attitudes toward government-imposed regulations. Life circumstances vary, as does the context of the need to conserve in different energy sectors: transportation, residential, commercial, public. Since each researcher has specific, albeit very different, definitions of the relevant context, it is not surprising that the results have been less than cumulative. The one conclusion that can be confidently drawn from the survey research is that the public is confused about what actions they should take; whether these actions will benefit them directly or serve some vaguely defined national goal of reducing total demand; and whether the actions they are asked to take are likely to make a real difference (to them or to the national goal).

The need for energy conservation resulting from energy shortages is relatively new to Americans; however, controversy surrounding the use of nuclear technology for peaceful purposes has existed since the late 1940s. Since that time, numerous surveys commissioned by government, industry, and private groups have been conducted on public attitudes toward nuclear electric power.[12]

Various wordings have appeared on surveys used to assess public attitudes toward nuclear electric power: general feelings or opinions about nuclear electric power; whether it should be used to generate electricity; whether (more) nuclear electric power plants should be built (in general, and in order to "meet the energy crisis"); whether nuclear electric power plants may be built "nearby" to where the respondent lives; and after Three Mile Island, whether respondents favor or oppose measures to restrict nuclear electric power (permanent closure of all plants, reduced operation or short-term plant closure, or the operation of only those plants currently on-line).

Figure 19.1 portrays a schematic representation of the trend in attitudes toward nuclear electric power from the 1950s through the present. It is *not* a plot of data; rather, the figure graphically displays a summary of many data points in order to present a rough shape of the trend. Until the mid-1970s, the data were relatively stable with only minor fluctuations in levels of support of or opposition toward nuclear electric power. Although some differences in the data may be observed, the relative stability of findings holds across almost all variations of question wording and context. Throughout this period there was a weak pronuclear electric power consensus: 50–60 percent in favor; 25–35 percent opposed; and 15–20 percent undecided, unsure, or with no opinion. For all surveys prior to the Three Mile Island accident, support levels were 10–20 percent higher than opposition levels.

Can a single set of events drastically alter public opinion? Dramatic events during 1973–78 produced noticeable, yet relatively minor, perturbations in the trends of opinion on energy matters; the major accident in early April 1979 at Three Mile Island, near Harrisburg, Pennsylvania, had very different results. Confusion was widespread: Was an evacuation necessary? Why was there no evacuation plan? Would a gas bubble cause an explosion? What was the cause of the accident? Who was in charge—the president, the Nuclear Regulatory Commission, the governor? Had radiation already leaked, and how much? Even after the crisis phase had ended, these and many more questions remained unanswered.[13]

Media coverage of the event was both intensive and extensive, and continued over a period of several weeks:

> The Three Mile Island accident was obviously a newsworthy event, yet the quantity of coverage given to it far surpassed what was given two months later to the nation's worst airline accident, the crash of a DC-10 that killed 274 people. Among the factors contributing to the inordinate quantity of coverage were these: The power plant accident occured at a time of public concern with nuclear power, so media people were sensitized to the issue and easily attracted to the event. By a quirk of timing, the antinuclear movie *The China Syndrome* was released to theaters across the nation just days before the accident. The event itself had great human interest—the week-long struggle with the bubble, the heroics of the nuclear engineers in averting disaster, the entry of the President, and the exit of the frightened populace—all of this had the drama of a soap opera, to be followed day after day. In addition, the plant site in Pennsylvania was easily accessible to reporters in nearby New York and Washington.[14]

Survey data on attitudes toward nuclear electric power collected after April of 1979 clearly indicate that the accident at Three Mile Island affected the public. For one thing, the patterns of attitudes became unstable and irregular. In fact, for the first time, opposition levels exceeded support levels in several surveys. On average, support levels decreased, opposition levels nearly doubled, and the level of uncertainty was cut in half.

**FIGURE 19.1   Schematic Representation of Trends in
American Attitudes toward Nuclear Electric Power**

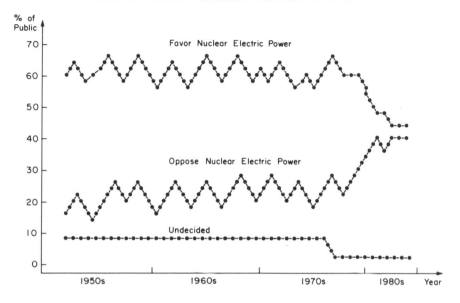

Wording of one type of question—whether the respondent would be for or against building a nuclear electric power plant "in this area" (within five miles of his or her community)—always elicited opposition levels somewhat higher than other wordings. Moreover, the decline in support for building nuclear electric power plants "in this area" began several years earlier than for alternative questions on more general attitudes toward nuclear electric power.[15] Until the mid-1970s public support for local nuclear electric power plant construction outweighed opposition by more than two to one. However, even prior to Three Mile Island, this attitude slowly but steadily changed to the point that the situation is now reversed: opposition levels are twice as high as support levels.

It is impossible to predict the direction these trends will take in the future. It is also difficult to assess the role of the media or the effect of public knowledge (or ignorance) about nuclear electric power on attitudes toward it. However, Allan Mazur concludes the following based on his analysis of media influences on public attitudes:

> [There is] the long-standing belief of nuclear proponents that public opposition would fall away "if only the public knew the facts." This belief persists despite abundant research showing that attitudes toward nu-

clear power have little relationship to knowledge about it (Mazur, 1981a). But worse, for the proponents' point of view, is this: the more that they attempt to fill the media with stories about the low risks and high benefits of nuclear power, the more the media are bound to carry the opposition story as well. And the consequence of this heightened media coverage—even though it might be balanced or slightly pro-nuclear— is that public opposition will increase, not decrease. If the proponents of nuclear power wish to serve their own interests, the evidence suggests, the less said the better.[16]

THE PERCEPTION OF RISK ASSOCIATED WITH NUCLEAR ELECTRIC POWER

Disaggregation of the data of pro- and anti-nuclear electric power attitudes produces an interesting finding. Those who favor this technology have relatively weak, heterogeneous beliefs about it. Of the plurality who are called supporters, few (about 25 percent) state that they strongly favor nuclear electric power. The advantages attributed to the technology are disparate and poorly articulated: provides cheaper power; does not pollute air and water; stands for progress; increases scientific knowledge; is efficient; conserves natural resources; reduces dependence on foreign oil; creates domestic availability.[17]

On the other hand, nuclear electric power opponents have a very homogeneous attitude structure and stronger behavior motivation.[18] Those who state opposition to this form of generating electricity are much more likely to feel strongly about their attitude than their supporter counterparts. The most serious concerns expressed relate to risks associated with the safety of plants, health, and environmental damage. These concerns are consistently mentioned and are clearly articulated, often with deep emotion: safety problems or risks; possibility of radioactivity leakage; problem of disposal of radioactive wastes; potential for accidents, explosions, or meltdowns; health effects; threat of sabotage; nuclear proliferation and terrorism; public anxiety.[19]

Researchers have been investigating the perception of risk associated with technology in general, and the risks associated with nuclear electric power have played an important role in that research. Studies conducted by psychologists using small groups in laboratory settings, rather than surveys, have determined that nuclear electric power is perceived to be a technology that provides benefits that are quite low and risks that are unacceptably great.[20] A number of qualitive characteristics of the technology are uncovered in order to elaborate on these feelings of perceived risks. Of these, nuclear electric power evokes greater feelings of dread than almost any other technology. In addition, nuclear electric power is viewed as a technology with risks that are uncontrollable, lethal, and potentially catastrophic.

The view held by nuclear electric power opponents that a serious reactor accident potentially resulting in thousands, even millions, of deaths and causing severe, irreparable environmental damage over a vast geographical area is likely to occur within their lifetime, contrasts sharply with the proponents' view that multiple safety systems will prevent or at least limit damage in the extremely unlikely event of a major accident. Some have argued that the opponents' perceptions are based on a lack of knowledge and are, therefore, "irrational."[21]

In order to extend the research on technological risk perception beyond the limited generalizability of the small-group laboratory research design, a study utilizing survey techniques would be required. Such a study was recently conducted: it compared the perceptions of technology held by elite groups of advocates with perceptions of the general population. More specifically, the study measured perceptions of technology's positive and negative consequences; perceptions of safety standards set to control technology in order to make it more acceptable; and personal involvement in activities (for and against) specifying the conditions for developing, maintaining, or utilizing technological products. Perceptions of nuclear electric power were examined along with five other specific technology referents: air travel, automobile travel, nuclear weapons, industrial chemicals, and handguns.[22]

Residents from two contrasting regions of the U.S. were sampled (Connecticut in the Northeast and Arizona in the Sunbelt, or Southwest). Although the sample design included specially selected pro- and anti-technological advocates from each region (n = 299), the following analyses report results only from the full probability samples of the general population (n = 1,021).

Respondents were asked to rate the risks associated with each of the six technologies studied on a scale from one (no risks) to seven (very great risks). In addition to overall risks, four qualitative characterizations of the risks were similarly rated for each technology (see Appendix for question wording for all items referenced from this study).

Table 19.4 displays the aggregate findings for the general population samples combined. The table reports arithmetic mean responses for each technology. The ratings were measured for each technology separately on the seven-point scale—they are *not* a rank ordering of the technologies by the respondents. The mean overall risk associated with nuclear electric power (5.0) was lower than the overall perceived risks associated with nuclear weapons, industrial chemicals, and even handguns. Of the four qualitative risk characterizations for nuclear electric power, its potential for catastrophe was rated the highest (4.7), although the perceived catastrophic potentials of nuclear weapons (6.4) and air travel (5.9) were rated greater. Dread did *not* appear to be an especially dominant factor in the minds of the respondents who rated the risks associated with nuclear electric power (mean = 4.0).

Similar ratings were obtained on seven-point scales for the restrictions and standards that apply to each technology. The scales of the level of regulatory strictness are composed of two parts, each rated independently for each technology: first the perception of current restrictions and standards, and then the perception of desired restrictions and standards. Nuclear electric power was rated on average at a moderate level of strictness (4.0), but respondents desired restrictions and standards that are extremely strict (6.6). The desired levels of regulation were higher than the perceived current levels for each of the six technologies measured (see Table 19.5).

An acceptability index was constructed to gauge the difference between current and desired standards. A positive number indicates current standards are stricter than desired; zero indicates perfect correspondence between the current rating and what standards should be; a negative number reflects a desire for restrictions and standards that are stricter than current regulations. All six technologies were perceived to be currently regulated less strictly than respondents felt they should be. Nuclear electric power's index value is neither high nor low compared with the other six technologies—less acceptable than air and automobile travel but more acceptable than industrial chemicals and handguns. The index value for air travel (the most acceptable) is a function of a perception that it is currently regulated at a fairly strict level. The least acceptable technology (industrial chemicals) suffers from a perception of currently being regulated at a very low level of strictness.

Do these perceptions of technology's risks relate to behavioral responses? Are perceptions of regulatory strictness—and notions about the acceptability of restrictions and standards designed to control technology's potential negative effects—associated with becoming personally involved to influence risk-management decision making? How widespread is pro- or anti-nuclear electric power activism? Indicators to measure the actions taken by respondents were developed: the items are ten self-reported activities that range from mild expressions of approval or rejection (e.g., letter to an editor) to forceful interventions (e.g., public demonstrations). All ten items were repeated six times—separately for each technology—and were preceded by a reference to the respondents' view about the restrictions and standards that apply to each technology (see Appendix).

Nearly a quarter of those interviewed had engaged in at least one of the ten activities in order to express their views on the perceived regula-

TABLE 19.4 Mean Scores of Four Risk-Perception Variables for Six Selected Technologies: Total General Population Samples from the Northeast and the Southwest

Perception of Risk Variables	Air Travel	Automobile Travel	Industrial Chemicals	Handguns	Nuclear Weapons	Nuclear Electric Power
Mean Risk: Overall	3.9	4.8	5.5	5.6	6.1	5.0
Mean Risk: Known to Scientists	2.7	2.4	4.3	2.6	4.0	4.1
Mean Risk: Fatality Estimates	3.4	6.0	4.1	5.5	2.8	2.7
Mean Risk: Catastrophic Potential	5.6	3.2	4.4	2.9	6.4	4.7
Mean Risk: Dread	3.0	2.5	4.1	4.5	5.5	4.0

Note: Original scale for each variable ranged from 1 (no risk) to 7 (very great risk); n = 1,021.

245

TABLE 19.5 Acceptability of the Restrictions and Standards of Six Selected Technologies: Total General Population (n = 1,021)

Restrictions and Standard Variables	Air Travel	Automobile Travel	Industrial Chemicals	Handguns	Nuclear Weapons	Nuclear Electric Power
Mean Current Restrictions and Standards	4.7	3.7	2.7	2.7	4.2	4.0
Mean Desired Restrictions and Standards	6.4	5.8	6.6	5.8	6.7	6.6
Mean Acceptability Index	−1.7	−2.0	−3.9	−3.1	−2.5	−2.6

Note: Acceptability index is the subtraction of the scale of "current restrictions and standards" from the scale of "desired restrictions," both of which were rating scales ranging from 1 (not very strict) to 7 (extremely strict). A positive number indicates current standards are stricter than desired; zero indicates perfect correspondence between the current rating and what standards should be; a negative number reflects a desire for restrictions and standards that are stricter than current regulations.

tory strictness that applied to nuclear electric power (Table 19.6). This level of action was the highest among the six technologies studied, although handguns and nuclear weapons produced similar levels of personal involvement. This finding should not be misinterpreted. It does *not* mean that anything like a quarter of the general population is engaged in highly emotional antinuclear activism for three reasons. First, the action question asks the respondents if they have *ever* engaged in these activities. Second, the measure includes not only actions against the technology but also actions in support of it. Third, some of the activities are clearly low-level, established forms of citizen participation that do not necessarily presuppose highly charged passion (e.g., signing a petition).

Of the 23 percent who reported having done at least one of the ten activities with reference to nuclear electric power, on the average two activities were undertaken. Nuclear electric power is the highest (tied with automobile travel) for this measure of the intensity of action; yet this dispels the common notion that activism is cumulative. Personal involvement of the general population is limited in scope and intensity. Relatively few people become involved at all, and even fewer become involved deeply enough to intervene intensively as measured by their self-reported actions.

For each technology under investigation, alternative policy tradeoff questions were devised in order to reflect the complexity and idiosyncrasies of the policy issues associated with them. The questions on regulatory strictness, reported above, allowed respondents to rate the desired level of regulatory control without having to consider the costs (social and economic) associated with stricter standards. The alternative policy tradeoff question presented balanced policy options, each of which contrasted with the others; the respondent was forced to consider the effects of choosing the option favored, since its alternative was incorporated in the design of the question. The tradeoff question for nuclear electric power presented three alternative policies concerning future expansion: build more as needed; continue operating those already on-line with careful monitoring, but build no more; shut down all plants permanently and build no more (see Appendix).

Figure 19.2 shows the data on the percentage who acted, for each technology, disaggregated by the alternative policy tradeoff questions. Of the total population, 23 percent acted in order to express their views about the restrictions and standards that apply to nuclear electric power: 5.4 percent acted and held the pro-technology view (build more plants if needed); 3.7 percent acted and chose the extreme pro-safety option (shut down all plants permanently). Of all the nuclear electric power activists found in the general population, slightly more took the pro-technology view than the pro-safety view on policy options dealing with

TABLE 19.6 A Profile of Action for Six Selected Technologies: Total General Population and Activism Subgroups

Action Variables	Air Travel	Automobile Travel	Industrial Chemicals	Handguns	Nuclear Weapons	Nuclear Electric Power
Percentage Who Acted (N = 1,021)	5.0	19.0	15.1	22.8	21.6	23.3
Mean Number of Actions by Those Who Acted	1.4	2.0	1.8	1.7	1.9	2.0

Note: For each technology, respondents were asked to indicate whether they had engaged in ten specific activities (plus any others) in order to express their view on the restrictions and standards that apply to the safety of that technology.

FIGURE 19.2 The Percentage Who Acted, by Alternative Policy Tradeoff Questions

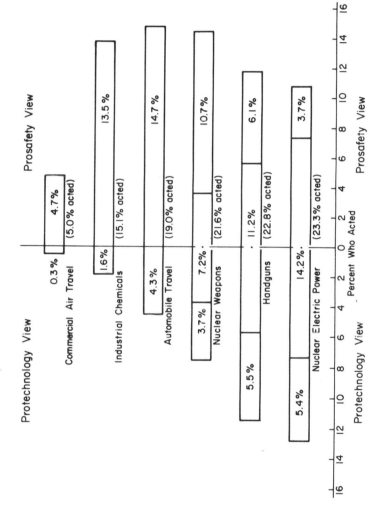

future expansion. The pattern for handguns is similar to nuclear electric power. For the other four technologies studied, a large majority of those who acted held the pro-safety view.

The analyses of survey data on energy issues presented above are intended to demonstrate the variety of information that may be obtained, beyond the headline poll, from this research design. When longitudinal data are available, an examination of changing trends adds perspective. When different surveys use different wordings, the effect of the various wordings on results adds depth to interpretations. When multiple questions on a given topic have been included in a survey questionnaire, additional dimensions may be included in the analysis. The information provided is not only more accurate, it is also more useful.

Little consensus of opinion is to be found in attitudes on energy-related issues. But the energy decisions to be made are political ones, and the role of citizen participation in a democratic political process will, as always, be balanced against other institutional mechanisms developed to resolve those issues. The results from surveys are not intended to play a direct role in that political process. Rather, they provide useful information for elevating the level of the debate.

NOTES

1. Al Gollin, "Adding a Few Strands to the Network," *AAPOR News* 12 (1984) (Princeton, NJ: American Association for Public Opinion Research), p. 7.

2. Albert H. Cantril, ed., *Polling on the Issues* (Cabin John, MD: Seven Locks Press, 1980), p.v.

3. Peter H. Rossi, James D. Wright, and Andy B. Anderson, *Handbook of Survey Research* (New York: Academic Press), 1983. See also Pamela L. Alreck and Robert B. Settle, *The Survey Research Handbook* (Homewood, IL: Richard D. Irwin, 1985).

4. Rossi et al., *Handbook*; Howard Schuman and Stanley Presser, *Questions and Answers in Attitude Surveys: Experiments on Question Form, Wording, and Context* (New York: Academic Press, 1981).

5. Herbert H. Hyman, *Secondary Analysis of Sample Surveys: Principles, Procedures, and Potentialities* (New York: John Wiley and Sons, 1972); Schuman and Presser, *Questions and Answers*.

6. Cantrill, *Polling on the Issues*, pp. 197–98.

7. Daniel Yankelovich, Robert Kingston, and Gerlad Garvey, eds, *Voter Options on Nuclear Arms Policy* (New York: Public Agenda Foundation, 1984), pp. 20–21.

8. Robert C. Mitchell, "Polling on Nuclear Power: A Critique of the Polls after Three Mile Island," in *Polling on the Issues*, ed. Cantril, pp. 70–71.

9. Barbara Farhar et al., *Public Opinion about Energy: A Literature Review*, report #SERI/TR-53-155 (Golden, CO: Solar Energy Research Institute), pp. 67–76.

10. William R. Burch, Jr. and Donald R. DeLuca, *Measuring the Impact of Natural Resource Policies* (Albuquerque: University of New Mexico Press, 1984), p. 163.

11. Ibid., pp. 163–64.

12. Ibid., chap. 10; Barbara D. Melber, Stanley M. Nealey, Joy Hammersla, and William L. Rankin, *Nuclear Power and the Public: Analysis of Collected Survey Research* (Seattle, WA: Battelle Human Affairs Research Centers, 1977); William R. Freudenburg and Eugene A. Rosa, eds, *Public Reactions to Nuclear Power: Are There Critical Masses?* (Boulder, CO: Westview Press, 1984); Farhar et al., *Public Opinion about Energy;* Mitchell, "Polling on Nuclear Power"; William L. Rankin, Barbara D. Melber, Thomas D. Overcast, and Stanley M. Nealey, *Nuclear Power and the Public: An Update of Collected Survey Research on Nuclear Power* (Seattle, WA: Battelle Human Affairs Research Centers, 1981).

13. Burch and DeLuca, *Measuring the Social Impact,* p. 168.

14. Allan Mazur, "Media Influences on Public Attitudes toward Nuclear Power," in *Public Reactions to Nuclear Power: Are There Critical Masses?,* ed. Freudenburg and Rosa, p. 101.

15. Freudenburg and Rosa, *Public Reactions to Nuclear Power,* pp. 53–55.

16. Mazur, "Media Influences," pp. 111–12.

17. Farhar et al., *Public Opinion about Energy.*

18. Ortwin Renn, *Man, Technology, und Risk: A Study of Intuitive Risk Assessment and Attitudes toward Nuclear Energy* (Julich, Germany: Kernoforschungsanlage, 1981).

19. Farhar et al., *Public Opinion about Energy.*

20. Fischhoff, Baruch, et al., "How Safe Is Safe Enough? A Psychometric Study of Attitudes toward Technological Risks and Benefits," *Policy Sciences* 9 (1978):127–52.

21. Freudenburg and Rosa, *Public Reactions to Nuclear Power,* chap. 6.

22. Donald R. DeLuca, Jan A. J. Stolwijk, and Wendy Horowitz, "Public Perceptions of Technological Risks: A Methodological Study," in *Contemporary Issues in Risk Analysis,* ed. Vincent T. Covello, Jeryl Mumpower, and Joshua Menkes (New York: Plenum, in press).

APPENDIX

Questionnaire Items from the Risk Perception Study
1. Overall risk perception and four qualitative risk characterizations.

> 46. All activities involve some risks. Accidents can occur no matter what measures are taken to avoid them. Think about the hazards and risks related to industries and their products: for example, risks of illness, injury or death to those who use the products, as well as those who make them; air, water, land pollution or other environmental damage; community disorder; national disaster. Consider **only** risks, not benefits. Think of hazards or risks to you and your family, as well as risks to the rest of the people in the country.

> A. Here is a scale, where "1" means **no risks overall** and "7" means **very great risks overall**, how would you rate each industry or product?

Also asked for:

> B. The risks of some industries and their products are **well** known and understood b~ scientists and technical people. The risks of other industries and products are **not we** known and understood by scientists and technical people. How would you rate eac~ industry or product on this scale?

> C. One of the risks of industries and their products is the risk of death. On this scale where "1" means **few** and "7" means **many**, how many deaths are likely to occur in this country in the next year, as a result of each of the following?

> D. The risks of death from some industries and their products affect people only one at a time. The risks of death from other industries and their products can affect large numbers of people in a single event. How would you rate the risk of death from each industry or product on this scale?

> E. Some industries and their products produce risks that people have learned to live with and can think about reasonably calmly. Other industries and products produce risks for which people have very great dread. How do you feel about each industry or product on this scale?

2. Perceptions of restrictions and standards (current and desired).

3. Now, I am going to ask you two questions about several technologies or industries. The first asks your opinion about current conditions—as they are **NOW**. The second question asks for your opinion on what the conditions **SHOULD BE**.

Some people believe that the current restrictions and standards that deal with **the safety of automobile travel** are not very strict—point number 1 on this scale.

Also asked for:

4. Some people believe that the current restrictions and standards that deal with **transporting and disposing of industrial chemicals** are not very strict—point number 1.

5. Next, think about the safeguards associated with maintaining **nuclear weapons** as a part of our national defense.

6. Next, think about the restrictions that deal with buying, selling, and owning **handguns**.

7. Next, consider the restrictions and standards that deal with the safety of **nuclear electric power**.

8. Finally on this question, think about the restrictions and standards that deal with the **safety of commercial air travel**.

3. Self-reported actions.

9. Some people do different things to make their feelings known on issues that concern them.

Here is a list of actions people have taken in order to express their views.

READ EACH ITEM.

HAND
CARD
C

Have you done any of these things to express your views on the restrictions and standards that apply to **the safety of automobile travel?**

Yes 1
No .... (GO to Q. 10) 2
Don't know or can't remember 8

	YES	NO	DON'T KNOW
A. Written a **letter, telephoned,** or sent a **telegram** to an editor, public official or company.	1	2	8
B. **Signed** a petition.	1	2	8
C. **Circulated** a petition.	1	2	8
D. **Voted** for or against a candidate for public office in part because of his or her position on this issue.	1	2	8
E. Attended a **public hearing** or a meeting of a **special interest organization.**	1	2	8
F. **Spoken** at a public hearing or forum.	1	2	8

254

G. **Boycotted** a company.

H. **Joined** or **contributed** money to an organization.

I. Attended a **public demonstration**.

J. Participated in a **lawsuit**.

K. Other:
Specify: _____

1 2 8
1 2 8
1 2 8
1 2 8

Also asked for:

10. Have you done any of these things to express your views on the restrictions that apply to **handguns**.

11. Have you done any of these things to express your views on the restrictions and standards that apply to **the safety of nuclear electric power**?

12. Have you done any of these things to express your views on the restrictions and standards that apply to **the safety of commercial air travel**?

13. Have you done any of these things to express your views on the safeguards associated with maintaining **nuclear weapons** as part of our national defense?

14. Have you done any of these things to express your views on the restrictions and standards that apply to **transporting and disposing of industrial chemicals**?

4. Alternative policy tradeoff questions.

30. Do you **favor** further increasing the safety of auto travel through such things as improved auto design and added safety features **or** do you **oppose** such measures because they would be too expensive?

Favor 1
Oppose 2
Not sure 8

31. Which **one** of the following positions comes closest to expressing your views on industrial chemicals?

READ TWO POSITIONS.

| HAND CARD E |

Position (1)   Industrial chemicals have contributed so much to our standard of living that we should do all we can to encourage the development and distribution of new industrial chemicals.

Position (2)   Industrial chemicals have been such a mixed blessing for our society (country) that we should be much more careful before allowing new industrial chemicals to be developed and distributed.

INDICATE POSITION TAKEN.

Favors Position (1) 1
Favors Position (2) 2
Not sure 8

32. Do you **favor** further increasing the safety of commercial air travel through improved airplane design and added safety features at airports **or** do you **oppose** such measures because they would be too expensive?

Favor 1
Oppose 2
Not sure 8

33. I am going to read to you three statements, each of which represents how some people feel about nuclear electric power generation. Please listen carefully and tell me which **one** view you favor the most.

READ THE THREE POSITIONS.

Position (1)   All currently operating nuclear electric power plants should continue operating, with careful safety monitoring. If more electricity is needed, more nuclear power plants should be built.

Position (2)   All currently operating nuclear electric power plants should continue operating, with careful safety monitoring, but **no more** new nuclear power plants should be built until more is known about the safety risks involved. If more electricity is need, some other form of power generation should be used.

Position (3)   All nuclear electric power plants should be shut down permanently and no more should be allowed to be built. If more electricity is needed, some other form of power generation should be used.

INDICATE POSITION TAKEN.

Favor Position (1) "continue operating" 1
Favor Position (2) "continued, but no new ones" ... 2
Favor Position (3) "shut down permanently" 3
Not sure 8

257

34. Which **one** of the following three positions comes closest to expressing your views on handgun restrictions?

READ THREE POSITIONS.

Position (1)   The use, sale, and ownership of handguns should be banned.

Position (2)   The use, sale, and ownership of handguns should be allowed, but should be restricted by licensing handgun dealers and owners.

Position (3)   The use, sale, and ownership of handguns for lawful purposes should not be restricted by government.

INDICATE POSITION TAKEN.

Favors Position (1) "ban"	1
Favors Position (2) "licensing"	2
Favors Position (3) "no restrictions"	3
Not sure	8

HAND CARD G

35. In order to maintain our national defense, which **one** of these three views do you favor most?

READ THREE POSITIONS.

Position (1)   We should strive to maintain nuclear superiority over the Soviet Union by continuing to manufacture nuclear weapons.

Position (2)   We should increase our arsenal of nuclear weapons to achieve equality with the Soviet Union.

HAND CARD H

Position (3)   Our current nuclear arsenal is more than sufficient to deter any potential aggressor; we need manufacture no more nuclear weapons.

INDICATE POSITION TAKEN.

Favors Position (1) 1
Favors Position (2) 2
Favors Position (3) 3
Not sure 8

37 Suppose two friends were having a discussion about how to deal with industrial chemical waste disposal. Both agree that there is the need to regulate the transportation and disposal of industrial chemical wastes and that current dumps should be cleaned up. However, each has a different opinion on paying victims who have been harmed.

HAND CARD J

Person (1):   feels the matter of compensating (paying) victims who have been harmed **should be settled on a case by case basis in the courts**. This means the victims would have to sue the company involved. Payment would result if the court found the company responsible for the harm.

Person (2):   feels the compensating (paying) of victims who have been harmed **should be regulated and paid for by the government**. This means the victims would file a claim with a government agency which, on finding harm done, could award payment from a fund contributed to by all the companies in the business. No single company would have to be proven responsible.

With which person do you most agree?

Person (1) .... "courts" 1
Person (2) .... "government fund" 2
Don't know 8

259

20

A New York State
Consumer Energy Mindset ————

Steve Barnett

Are most consumers ignorant of energy facts, or are the utility companies trying to con consumers? This chapter will suggest that neither position reflects the way New York State consumers actually think about their daily energy use. Most residents of New York have a consistent, considered perspective (which we will call a "mindset") on energy, including attitudes about fuel choice, utility operation, role of government, and conservation. Scientists and engineers can be found who would agree or disagree with each aspect of this mindset, but the important point is that the views of such scientists and engineers are not likely to change the minds of many ordinary consumers.

We will develop the essential features of the New York State energy mindset based on a series of group interviews with 64 consumers around the state. These interviews allowed participants to talk freely among themselves (for about two-and-a-half hours) about ther feelings on energy. These interviews were then transcribed. They were analyzed, using anthropological methods, to understand decision-making styles, key symbols and phrases, and underlying dimensions of meaning. In addition, we have surveyed approximately 450 New York State residents as part of national and regional studies.

MINDSET COMPONENTS

The first issue, of course, is utility rates. Oddly enough, although people will talk about rates with intensity, the importance of high rates for defining attitudes toward utilities is declining, except where nuclear plants may be coming on-line. Rates are not as important as they were

4-5 years ago. What now concerns middleclass customers about electricity rates across New York State is a frustration with conservation efforts (perhaps even a backlash), and questions about the quality of utility management. People perceive that they have taken steps to conserve, but many do not perceive a benefit derived from these actions (in part, because utility management will "just raise rates"); therefore there is a tendency to resist taking further steps and a reluctance to continue current levels of conservation activity.

Another effect of perceived high rates is that for poor people, especially the urban poor in New York State, there is an increasing sense of a split between usage and bill size—most urban poor people in New York State do not think that their bills accurately reflect usage. Many think the utility decides how much revenue it needs and then bills accordingly, quite apart from actual usage. Lower-income black customers tend to believe there is a separate and higher rate for blacks.

The next key attitude in this complex utility and energy mindset concerns the legitimacy of accountability procedures and regulation. The Public Service Commission (PSC) is now one of the institutions that give rise to a general public suspicion about the accountability of utility companies' bill size, bill components, and management quality. In fact, suspicion of the equity or fairness of the PSC is the latest contributor to rate "approximation" in New York State customers' minds. The first approximation was, "utilities are just making too much profit, that's why my rates are so high." The second is, "They may not be making that much profit but they certainly are mismanaged, that's why my rates are high." The third approximation is, "Maybe they're making too much profit, and maybe they're mismanaged, but surely the Commission is in their hip pocket." Considering these three approximations as cumulative, it is no wonder that the ordinary citizen quickly becomes suspicious and cynical of the entire utility billing and rate-making process.

The third mindset component stems from a feeling of loss of control on the part of the individual customer: specific conservation efforts seem pointless; the fairness of the PSC is in question; the ability to think about or alter their energy future is undermined. This is especially true for customers of New York State utilities with nuclear power plants under construction or nuclear power plants on-line.

Another key component of how New York consumers think about utilities results from their understanding of utility and energy economies. Frustrated utility managers often bemoan the lack of customers' economic "education," but as an anthropologist I would suggest that there is a well-developed "native economic" theory in the United States and in New York State as well, and that this native theory conflicts at important points with the daily operation of the electric utility industry:

- Most people in New York State think that there ought to be a discount for volume; the price of any commodity purchased in bulk ought to be discounted. Some utilities have summer rates—that is, when more electricity is being used, rates increase. The reasons have to do with load management and relatively inefficient "peakers," but these conflict with a basic notion of native economics and such concepts are not easily changed.
- Most people think that excess supply should lower price, the classic, if simplistic, idea of a supply/demand relationship. For many utilities, excess capacity results in higher costs, given idle plant and fixed costs.
- Many people, when they purchase a house, have a "mortgage rationale"— they know that total payment for a house may be higher with a long-term mortgate than if they could pay a large sum up front, but they are willing to incur this larger total payment in exchange for smaller monthly payments. Most utilities think that if customers understood that the Cost-of-Work-in-Progress (CWIP) in a rate base would equal a smaller overall price for a plant, they would approve. In fact, customers using their own native economic rationale can grant that view and still want a lower monthly payment.
- Most people in New York State, and most people throughout the country, believe they have a right to electricity. "Right to electricity" is a complex phrase; here I am taking it to mean that a consumer should have a minimum amount of electricity regardless of his or her ability to pay. This conflicts with the utility industry's attempt to send customers price signals in various ways. Time-of-day rates have not worked well, and many customers think that such rates are an inappropriate use of an economic sledgehammer.

The next component of this overall New York mindset is the strong media influence on public opinion in this state, primarily through newspapers and secondarily through local television news programs. For example, the continuing *Newsday* coverage of the Shoreham nuclear plant essentially defines what Long Island customers think about that plant. They repeat the language and the concepts of the *Newsday* material. The New York *Times*, for a larger and more elite constituency, both upstate and downstate, performs a similar function. Local television news, especially in Buffalo and New York City, is held in higher regard than is such news throughout the rest of the country.

Conservation is critical for an energy mindset. About 30 percent of New York State residents are in a group that feels what I will call "conservation backlash"; consumers know they have additional conservation measures to take but are not taking them because they do not perceive sufficient benefits from previous conservation activity. Lower-income consumers are taking fewer if any conservation measures. Typically, they refuse audits because of a perceived lack of resources to take advantage of audit recommendations.

The PSC has mandated that utilities throughout New York State spend 0.25 percent of total revenues in 1985 on conservation programs. These are oriented toward electricity consumption, and include commercial and industrial customers. I would caution against being too optimistic about the short-term results of such programs without careful rethinking of the nature and intent of conservation activities.

Consumer understanding of the fuels used to generate electricity is a critical part of any overall energy mindset. Most people in New York State do not think there is an electricity supply problem now or in the proximate future, despite (debatable) technical evidence to the contrary. Many consumers assume there is such a large supply available, that utilities just have to tap into it. Or they may agree that new plants will eventually have to be built, but in that case they also think that utilities have the resources and the finances to construct those plants. Most people in New York State, for example, think that utilities borrow at subsidized rates considerably below the prime rate. Hydro and nuclear are linked symbolically for many New York State consumers. Most people in New York State are aware of the relative abundance of hydro, more so for Niagara Falls, less so for Canadian hydro. Hydro is seen as having positive symbolic attributes; nuclear as having negative ones. Nuclear is unknown, dangerous, expensive, limited; hydro is known, safe, cheap, and renewable (unlimited). It is this juxtaposition of hydro and nuclear in New York that gives focus to much antinuclear sentiment in the state.

Let me very briefly discuss coal. Most consumers simply do not like it. When they say coal is "dirty," they are not primarily thinking about acid rain (a more elite concern) but, rather, coal's sooty, gritty quality. In New York State, we found that many people are concerned about the backwardness or the outdatedness of coal in terms of "competition with Japan." If New York State is using backward, outdated fuels, then somehow it is a backward, outdated state, and other countries are going to leapfrog over us.

Fears of an oil shortage or sharp price jump have declined. This result mirrors standard surveys and suggests a corresponding decline in the significance of an OPEC threat. To that extent, such a threat by itself cannot be used to justify all sorts of utility action (e.g., coal reconversion).

The new high-voltage transmission lines in New York State are emerging as a potential issue. This is true across the country. In New York, groups have formed in opposition to transmission lines that provide Canadian power. The focus of concern is safety, the use of pesticides along the right of way, and general ecological and aesthetic issues.

A separate study of 45 ordinary consumers around the state asked them to invent their own utility. They were instructed to work from scratch, forgetting about their private or public utility. Ignoring the specific situation you are in, they were asked, how would you design your own utility? The following are fundamental features that emerged in every one of those groups:

- Every group wanted much more local control; they wanted to compensate for a feeling of lack of control over energy costs and safety concerns. They resented utilities as abstract, faceless corporations; they wanted local officials to have more control over the decisions made by the utility.
- Every group also wanted smaller generating units. None of the participants had read Schumacher or Amory Lovins on changing economies of scale, but their theories have now percolated down to the popular culture, and many people believe that smaller generating units somehow result in more control, more flexibility and more efficiency. In fact, one group rejected the terms *plants* and *generating units* in favor of *modules*, a high-tech word.
- These modules, participants hoped, would run primarily on alternate energy forms (hydro, solar, wind, tides, etc.) with significant amounts of co-generation. We found in New York State a surprising number of people who were interested in residential co-generation—not simply commercial co-generation, but in each house having its own generator. As some of you may know, a number of companies, Japanese and American, now market a home generator that in some parts of the country can compete economically with what the utility is offering.
- Given a stress on local control and small generating modules, participants recognized the potential for supply problems, and so they simultaneously created a statewide, superefficient grid to shunt power around the state as needed.
- If participants felt the invented utility had sufficient local control, was using smaller generator units, and was linked up to a superefficient grid, they did not prefer that the utility be public. Under these conditions, a private utility was viewed as legitimate and more efficient, providing it operated in a "business-like" fashion. A criticism of private utilities in New York State is they they do not operate in a business-like fashion, and therefore profit is illegitimate; profit per se is not perceived as illegitimate.

We have identified the major components of an energy mindset. They include attitudes toward rates, utility management, governmental regulation, control of consumption (including conservation efforts); theories of native economics; role of the media; perceptions of nuclear, coal, and alternate energy; and possible alternatives to the present system. This complexity suggests that the mindset is well formed and can be taken for granted as the filter through which most New York State residents perceive energy issues.

What does this mindset imply for energy issues critical to the state? First, it is a relatively fixed perspective, for most of New York State citizens, about how the present utility system works and what they would like as an alternative. In other words, this mindset is not something easily changed. This suggests, for example, that further behavior modification in the direction of conservation will be difficult to achieve, given significant conservation backlash. We conducted an experiment where we changed the bathrooms of ten middleclass participants to resemble the bathroom of an equivalent middleclass household in Italy (complete with a small water heater in the bathroom and switches for its controller). Adaptation was grudging at best. Afterwards we administered group interviews where participants were asked how they felt about the experiment. One person said—and he was not exaggerating, he was just more pithy than the rest—"If I had to live that way permanently, I'd go to war." I doubt there will be much additional residential conservation behavior modification (as opposed to structural changes on account of new housing and appliance turnover) unless there are severe price shocks in the near future. Or, unless communities and towns are the focus of conservation incentives.

Examining supply, demand, and plant siting as three critical issues, the New York State public does not think supply is a problem, given available funds plus enough lead time. They also think that future technologies (as yet unspecified) will provide a fix-all. If you say to consumers, we have to plan now for the next 10–12 years, they tend to respond, "No, by that time, we'll have all sorts of innovative technology." And the planning horizon of most citizens of New York State is much smaller than 10–12 years, it is more like 3–5 years. Future residential supply is not a public concern, and I should add that most people in New York State think that there is enough supply for industry as well.

Demand and load management are more complex. I do not think we can anticipate that conservation will have the same moderating effect on residential demand that it has had in the recent past, given conservation backlash. New customers (ages 18–21)—and this is a consumer area hardly explored—are much freer in their use of energy. You may have noticed that Americans have moved back up to mid-size cars, with rather alarming speed. In general, younger people tend to use their discretionary income to achieve relatively quick gratification, and perhaps spend more money on energy than might have been anticipated. If electricity is powering devices younger customers see as necessary or desirable, they will allow a larger budget share for that electricity.

Finally, siting issues: these are critical in New York State. Coal conversion is a good example of an issue causing local resistance and concern. People do not necessarily see increased tax benefits as a decisive advantage provided by a conversion project. They are nervous about having a coal plant in their community. High-voltage transmission lines are an emerging siting issue that could become critical as opposition groups dig in to fight recent PSC approval of a 185-mile line running through central New York State.

CONCLUSION

Both government and utilities have to reckon with consumer attitudes, especially if those attitudes assess the regulatory process as weak and biased, and see utilities as poorly managed. Conservation programs, especially, must start by recognizing the potential for consumer resistance. Business-as-usual programs that do not address conservation backlash will continue with less than hoped-for results. Innovative conservation programs need to be developed that:

- stress vivid, personal information, not impersonal statistics;
- use community-based leaders and local organizations, not simply mass media messages;
- focus on what the consumer is losing now, not only invoke promised savings later;
- do not require capital outlay for low-income customers;
- create meaningful measures of success, avoiding confusing concepts of kilowatt-hours or therms.

New York State government needs to alter residents' impressions that:

- the PSC is ineffective;
- utility executives can outsmart governmental agencies and procedures;
- safety standards are lax, for coal, nuclear, and high-voltage transmission lines;
- alternative energy research and projects have been downgraded or, in some cases, abandoned;
- an imbalance exists between hydro-power allocated upstate and downstate.

Private utilities need to:

- recognize significant customer opposition;
- develop innovative rate structures with more customer choice and options;
- create real rewards and incentives for conservation;
- demonstrate efficiency of daily operation;
- acknowledge that lower-income customers often simply cannot pay and may have continuing difficulty paying, given present bill payment systems;
- avoid slick public relations campaigns that only further antagonize customers.

These are not easy tasks, to say the least. But their scope suggests the extent to which New York State residents are dismayed and confused by present policy and programs.

21

Predicting the Next Regulatory Crisis: An Issues Management Approach————

Howard Axelrod

To contend that a utility's management does not face the same business pressures as the unregulated world about it is to fail to respect, or at least recognize, the ability of regulators and professional interveners to influence what has traditionally been an exclusively managerial decision-making process. This process includes decisions by managers to expend billions of dollars on construction projects that will meet the energy needs of consumers for generations to come. The choice, to build or not, will affect the financial viability of the utility, as well as the economic foundation upon which many local communities and businesses build.

While some of the contemporary issues being raised by regulators and interveners representing consumers appear to be trivial, if not trite, many are based on rather sophisticated and complex economic, engineering, and accounting principles. For example, the application of marginal cost principles through time-of-use rates requires detailed engineering analysis of marginal production costs, in addition to an advanced understanding of microeconomic theory. Many issues are so complex that they pit one group of consumers against another, one region against its neighbor, as often as they pit consumer against utility. As much as we would like to think that all of a utility's decisions are based upon sound principles and accurate facts, often a utility is forced to circumvent political or regulatory barriers by the expediency of choosing short-term lower prices in exchange for long-term economic benefits.

If in reality utility management must address politically sensitive regulatory issues by convolving them into their decision-making process, then those issues that can affect the utility the most must be fully understood, if known, and predicted, if not. I do not mean to imply that the ultimate decision must be influenced by these exogenous forces, but that within the management process they must at least be evaluated to ascertain their potential impact.

Can regulatory issues be predicted? This chapter outlines one method of analyzing which issues, either generated by the regulator or by the consumer advocate, might reach crisis proportions in the foreseeable planning horizon. The possession of this knowledge provides management with the time and opportunity to consider those issues that appear to be sound, and delay or eliminate those of dubious nature.

REGULATORY INTERVENTION

Regulatory intervention comes in many forms, addresses many issues, and can affect a utility in a variety of ways. What is most important is to be able to isolate from the myriad of issues those that can affect the utility the most severely. Obviously, no single formula can categorically pinpoint which issues will reach crisis proportions next; however, there is a technique, presented here, that can help management identify those issues having the greatest potential to impose regulatory and/or political constraints upon them. Such a technique permits evaluation of the five basic regulatory concepts to be discussed and determines which are affected by the regulatory issues under review. It is assumed that the more concepts affected, the more likely it is that particular issues will sustain public attention.

BASIC REGULATORY CONCEPTS

The five regulatory concepts used to measure the likelihood that a particular regulatory issue will reach crisis proportions are:

- Equity
- Risk
- Incentives
- Economic impact
- Used and useful doctrine

Equity

The concept of equity requires that the electric prices charged to consumers be accurate and efficient, and that they properly allocate costs among the various consumer classes as well as among users of the same class of service over time. The current discussions surrounding the application of economic depreciation techniques versus the traditional accounting approach of straight line depreciation centers on intertemporal inequities. Because straight line depreciation actually declines in real terms over the lifespan of a capital investment, today's consumers are being charged relatively more for the same quantity and quality of electricity than future consumers will be.

Risk

Two important decisions by the United States Supreme Court[1] stipulated that a regulated utility be given reasonable opportunity to earn a rate of return at a level competitive enough to attract capital for future ventures. Stated another way, a utility, while not granted a guarantee, should be granted a return on equity no less than any other firm of similar risk. Webster's defines "risk" as the degree or probability of loss. As a result, a utility given a return greater than a risk-free government note must be prepared to assume some level of financial risk, namely some probability of loss.

Regulators must, in some circumstances, differentiate legitimate expenses from those that are imprudent or a result of mismanagement, in allocating costs between ratepayers and investors. Commissions must also consider changing market conditions and their effect on utility expenses, a risk investors must bear. Inflation can erode a utility's earnings until new rates are fixed. However, a basic regulatory principle concerning retroactive rate making prohibits the recovery from consumers of past losses in earnings.

Incentives

Conceptually, a regulatory body serves as a surrogate for the competitive marketplace. Unregulated firms adjust supply to match their product's free-market demand at some equilibrium price. A monopoly, unregulated, will maximize profit by offering its product at a relatively higher price than its competitive counterpart. This price will be met as supply is artificially restricted. In order to achieve the lowest prices at the greatest level of production, a regulatory body must be employed to fix "just and reasonable" rates. It is a key function of the regulator to provide incentives for efficient production that are not normally provided by an unregulated monopolistic market.

Economic Impact

As the price of electricity increases relative to other commodities, the consumer's response to price increases becomes more acute, or in economic terms, the consumer's willingness to buy becomes more price elastic. Traditional rate-making principles can at times distort or produce significant discontinuities in price patterns of electricity. For example, because most if not all construction expenses are accumulated without the provision for flow-through of cash earnings, the cost of large construction projects once entered into rate bases can produce major rate increases on the order of 50–100 percent. This type of hyperincrease, recently labeled "rate shock," can severely restrict a utility's total revenues as sales decline or growth diminishes. A utility's long-term growth opportunities must be measured against the economic impact of its pricing structure and competitive position.

Used and Useful

One of regulation's earliest concepts involved a definition of "utility plant" as both used and useful. Historically, many utilities began operation as subsidiaries to unregulated firms. Questions soon arose as to the legitimacy of or need for certain equipment and properties claimed to be required for the production of electricity. Regulators, in their attempt to derive accurate costs of service studies, limited inclusion in the rate base to those utility plants considered "used and useful."

This concept is today being applied quite differently to legitimize the existence of utility plants that are either in excess or not in operation.

RULE OF THREE

A review of many regulatory issues being proposed by consumer interveners reveals that the ones most ominous to the utility industry are those that touch upon at least three of the regulatory concepts outlined above. Table 21.1 illustrates this approach by providing an analysis of several contemporary regulatory issues under review throughout the nation. The range of issues presented represents some of the most popular concerns of consumer advocates and regulatory staffs. By no small coincidence, the issue of including "construction work in progress" in the rate base, an issue that is under intense public review, involves all five regulatory concepts. To the other extreme, management salaries and advertising expenses, issues often raised by consumer interveners but rarely acknowledged by the regulatory commission, involve only one of the listed regulatory concepts.

APPLICATIONS

The approach outlined above has two basic applications. First, critical regulatory issues can be ranked as to the likelihood of their reaching

TABLE 21.1 Regulatory Issues Matrix

Regulatory Issues	Equity	Risk Sharing	Incentives	Economic Impact	Used & Useful	Total
CWIP[a] in rate base	X	X	X	X	X	5
Excess capacity	X	X	X		X	4
Abandoned plant		X	X		X	3
Normalized accounting	X	X	X			3
Cost overruns (prudent)	X	X	X	X		4
Fuel adjustment clause		X	X			2
Rate shock	X	X		X		3
Management salaries			X			1
Advertising expenses			X			1

[a]Construction work in progress.

crisis proportions. For example, the fuel adjustment clause, which captured national attention after the codification of the Public Utility Regulatory Policy Act of 1978, is only being addressed on a limited basis, and in most instances only relatively minor adjustments are being made. Probably the most extreme changes are occurring in New York State, where the pass-through provision of one utility has been restricted to 80–90 percent of the deviation from base rates. Table 21.1 lists the two concepts affected.

However, the issue of flow-through versus normalized tax accounting affects three concepts that classify it as a potential crisis issue. Should utilities be required to flow-through all deferred tax benefits, as is being proposed by a number of consumer advocates and political representatives, billions of dollars would have to be raised to meet this obligation.

The second application of this technique involves the forecasting of crises. Once an issue is determined to be capable of creating a crisis, a review of the concepts actually being raised over time can help determine when the issue under review becomes critical. Actually, all that this technique is doing is crudely measuring the sophistication of consumer advocates' arguments before the regulatory body. The more basic the regulatory issues concepts being addressed, the more likely the commission will seriously consider adopting specific proposals.

CONCLUSION

The regulatory issues matrix in Table 21.1 can provide management with a handy tool for evaluating which of the regulatory issues being raised should be targeted for serious review and response. How management responds to these issues is not a subject of this chapter. However, the process by which each regulatory issue is analyzed against the five basic regulatory concepts will help the analyst determine a position on the subject, as well as the response strategy, be it adoption, rebuttal, or settlement.

NOTE

1. See *Bluefield Water Works and Improvement Corporation v. Public Service Commission of West Virginia*, U.S. Supreme Court 262 U.S. 679 (1923), and also *Federal Power Commission v. Hope Natural Gas*, U.S. Supreme Court 320 U.S. 591, 605 (1944).

PART VI
THE FINANCIAL
ENVIRONMENT————————————————

All sources of electricity supply and of conservation are capital intensive; in fact, electric utilities are two to three times more capital intensive than any other major nonregulated industry in the United States.[1] Therefore, investors must provide a substantial infusion of money simply to erect facilities to meet the modestly growing demands for electricity that authors in Part II predict. In fact, given that the current capital structure of most privately-owned regulated utilities contains a substantial portion (40–60 percent, typically) of permanent long-term debt, a persistent demand is placed upon financial markets as each series of long-term bonds matures and is refunded merely to sustain the current level of facilities. Over the previous decade, the combined demand for capital by electric and gas utilities has represented approximately 10 percent of all nonresidential investment in the U.S. economy.[2] The financial needs of this industry are obviously enormous and can have an impact on overall financial market conditions. Reciprocally, as a result of disproportionate construction cost escalation rates, in recent years the fraction of total electric rates attributable to capital costs has increased, so overall electric rates have become more sensitive to variations in rates of return required by financial markets.

As a result of the turbulent financial market conditions of the 1970s and the financial disasters caused by stalled nuclear construction projects, two questions are central to the industry's future. First, can the money be raised to sustain this industry, or will its growth be constrained by an inability to feed its voracious appetite for capital? Second, under what terms and at what price can this industry's needs for capital be met?

Despite the pessimism of many members of the financial community regarding the industry's future ability to raise money, we sense that many of the recent debates are really more of the "Is-the-glass-half-empty-or-half-full?" variety, rather than concerned with forecasting the level. The authors who address these ability to finance questions represent the following three major markets supporting private and public electric utilities: long-term markets for private utility securities (Hyman, Kelley, and Toole), long-term markets for public power agencies (Forbes), and short-term bank loans for construction (Kron). All of the authors in this section outline serious problems faced by the industry, and they indicate regulatory and institutional changes that may be required in order to enhance the industry's future ability to finance. Yet once the current nuclear facilities in New York (and elsewhere) are completed or disposed of, they seem to be saying, the electric utilities will be able to raise the money to meet their projected future needs (barring some unforeseen catastrophe affecting U.S. financial markets, or mismanagement). Nevertheless, these authors do call for changes in regulation and the way utilities plan for the future. They emphasize how important financial constraints are to the electricity industry—they can be binding.

NOTES

1. This comparison is true at the level of aggregation for industries by their three-digit Standard Industrial Classifications. See Richard Schuler and Robert Smiley, "Electric Utility Deregulation: Estimated Cost of Capital," Annual Meeting of the American Academy of Arts and Sciences (Detroit: May 31, 1983).

2. See Jane Gravelle, "Capital Stocks and Investment Flows in the U.S. Economy: The Effects of Economic Pricing in the Electric and Gas Utilities," *Resources and Energy* 7, no. 1 (March 1985):133–40.

22

Financial Constraints on Electric Utilities in New York State: Technology, Institutional Structure, and Investors' Attitudes

Leonard S. Hyman
Doris A. Kelley
Richard C. Toole

To begin with an aside, we would not be discussing New York State's electric future if the electric industry were not regulated. Most of the problems we face come about because of the peculiar nature of the regulatory marketplace, in which a few mortals try to stand in for the competitive marketplace, and a lot of others act as if the regulators either will or can do that job. It is touching to see grown people bet billions on the premise that the workings of the laws of supply and demand can be arrested. OPEC is one example of the folly of that attitude. Utility regulation is another.

Sometimes regulation—or lack of it—does succeed in making an imprint on the marketplace and participants therein. California regulators wanted alternatives to more, massive stations. They got what they wanted, and California utilities are not only different than before but more prosperous. Wisconsin regulators wanted tightly run utilities that eschewed big in favor of practical. Wisconsin utilities are among the most prosperous. Florida law demanded and got cost-effective conservation and reduced dependence on oil-fired generation. Texas was laissez-faire and pro-growth, and the utilities may not have set much store, as a result, in non-Texas-sized alternatives.

New York State's regulation has been guided by exceptionally sophisticated, talented people. The regulatory staff is highly professional. The state has an energy office and a power authority and one of the nation's most sophisticated state power pools. The state legislature has been relatively responsible in its relevant lawmaking so far. One would expect New York State's utilities, operating in such an environment, to be the industry leaders, exhibiting a uniform character of innovation, cost-effectiveness, and competence in a state whose agencies would tolerate nothing less. But, with one exception, that does not seem to have happened, at least in our perception—which is not to say that New York utilities are poorly run, but that they do not seem to be leading the pack despite the circumstances. We do not know why, or what went wrong. We do not know how things might have been different if the various governors had taken different stances, sooner, rather than later. We do suspect, though, that what has happened in New York would not have happened in Wisconsin as a result of regulation, and it would not have happened in several other places because, while regulators were not in control, utility managements were.

FINANCIAL STATUS

New York State's electric utilities are a varied lot, financially speaking. Consolidated Edison (ConEd) hasn't built anything in years. It is awash with cash. Orange and Rockland is strong, despite some non-nuclear spending projects. The Long Island Lighting Co. (LILCO) has been on the brink of bankruptcy for some time: It must either phase Shoreham into the rate base and/or write off a substantial portion of the Shoreham investment. The other four upstate utilities, all involved in Nine Mile Point, are in better shape than many utilities building nuclear power stations, but that does not say too much.

Whether the LILCO does or does not finish its nuclear plant at Shoreham is a political—not a financial—problem. Financial solutions would come if there were political ones. Our projections, which are based on incremental analyses that focus on factors that change net income and cash flow, indicate that once the political problems are settled, regulators could handle the plant without either bankrupting the LILCO or causing its customers to flee Long Island. Over a period of time, the LILCO could be rehabilitated financially and meet its service obligations to its customers and to the state power pool.

The Nine Mile Point utilities can, with the treatment we anticipate, handle the financing of that unit, although some utilities clearly come out ahead of others. Phase-in of the plant, both before and after completion, will limit rate shock.

IS THERE LIFE AFTER SHOREHAM AND NINE MILE POINT?

After the two nuclear facilities are completed and regulatory treatment is prescribed, then what happens? What happens depends on what electric customers need. If we are to believe the New York Power Pool's

(NYPP) projections, the New York State utilities will have a reserve margin in excess of 40 percent through the early 1990s and 20 percent in 2000. Admittedly, a lot of the capacity is oil-fired and much of it is old, but it is there to serve, and there would be plenty of it even if expectations for growth in demand doubled. There is time to act intelligently rather than hastily.

Considering the experience of the state's utilities in trying to serve customer growth through generation expansion, it seems safe to say that they will not be in a hurry to repeat the process or even to undertake heavy cost-reducing capital expenditures under the current system. Thus, capital expenditures could fall sharply after 1986. That would make phase-in more palatable to all and would lead to improved financial health for the utilities.

Such an outlook could fall prey either to unexpected and uncontrolled expansion of demand for electricity or to stringent acid rain legislation. In either case, capital expenditures would have to be raised, but at least in the former case, there would be some accompanying improvement in cash flow.

If the investor-owned utilities are allowed to build up a sufficient equity base—which seems unlikely, given the most recent ConEd rate decision—then they might be able to handle the next round of expensive power stations, which might have to be started by the early 1990s. Just keeping up with demand, as it is envisioned by the NYPP, could require one big power station every three years at a cost, to translate into dollar terms, of about 2–3-plus billion dollars every time. It may be do-able financially, with proper advance planning, but no one is sure where the stations will go. Perhaps the financial task involved in banking sites could be handled by the New York Power Authority (NYPA), so that utilities could act quickly when there is a need to build.

Of course we should consider the alternatives. The utilities certainly will, because they have learned a lot from Shoreham and Nine Mile, and not just engineering know-how, either. They learned that big power stations are risky and the regulatory process neither recognizes that nor shares risk and reward symmetrically. Thus, the most unimaginative approach, which may be the proper one for at least one of the utilities, will be to build a series of smaller (400–600 MW) plants in lieu of one large station. A well-built smaller plant, put up in four years, does not create the same financial strains or rate shock a large one does. If the companies concentrate on such units, we believe that the investor-owned utilities could easily meet the projected growth in load without financial strain or the need for extraordinary action on the part of either the state government or the NYPA. If the projections for growth prove too low, or if environmental controls on new facilities prove to be onerous, or if environmental controls placed on old facilities drain capital resources, then all bets are off. We doubt that investor-owned utilities would engage again in any programs that would jeopardize their financial well-being. Thus a program that would simultaneously involve power station construction and major pollution control expenditures would be avoided. In addition, the large power project is unlikely, because managements and investors will view such projects as something to avoid at all costs. A large power station is a lumpy investment: it can induce rate shock because of its size in relation to the old rate base, and it takes too long to build. While a large, jointly owned station can replace several smaller, individually owned stations, multi-owner projects have been notoriously unsuccessful. We do not expect utilities to go back in that direction. If it turns out that large projects are the least expensive alternatives and that they are really needed, we suspect that investor-owned utilities will gladly swallow their objections to socialism and let the NYPA do the building. Perhaps the NYPA perceives financial risk in a different manner.

ARE THERE ALTERNATIVES?

We believe that most discussions of utility marketing and finance are wrong-headed because they begin with the premise that the utility is in business to supply central station power within a natural monopoly. In reality, the utility's job (for which it should be paid) is to assure that customers can meet their needs for light, heat, machinery, and so on, in the most cost-effective manner. The most cost-effective manner could be central station electricity, or co-generation, or insulation, or microcomputers that regulate consumption. It might even be rehabilitation and better management of existing power stations. When the market is viewed in this manner, it becomes clearer that the utility does not have a monopoly and that there is no reason why it should not service customers through unregulated organizations. Nor is there a convincing case to retain the capital structures that go with the natural monopoly.

New York's utilities have done some work examining alternatives, but pushing alternatives now may not be cost-effective in the face of today's capacity forecast, as well as the availability of Canadian electricity. The real question is how to encourage the most aggressive development of these alternatives so that New York's consumers have an ample supply of economical energy and investors are willing to finance that supply at reasonable costs. Investors are likely to be predisposed to low-risk, small projects. How do these get built in a regulated environment? One solution would be to place all new expenditures that affect supply or demand in separate accounts or subsidiaries and either provide bonus returns for the most cost-effective projects, or effectively deregulate through avoided-cost pricing or through honestly deregulating. We see little incentive, under the current system, for utilities to do more than make minimal investments to keep the system going. There is not a lot of reason to expend capital to reduce costs if the consumer gets all the benefit—unless competition is so tough that cost reduction is necessary to retain the customer, in which case there is no need for regulation.

FINANCIAL AND CORPORATE RESTRUCTURE

Electric utilities are fully regulated, integrated entities. The bulk of the capitalization consists of mortgage debt, which means that property is set aside as security for the payment of the debt. The covenants that set the terms of the debt often have restrictions on disposition of property, expenditures and maintenance of property, and even on ability to pay dividends. At some time in the future, as new sources of power and new ways to serve customers are developed, the unitary, regulated structure of the utility might become a stumbling block, a deterrent to competitive response and to flexibility. But the massive issues of mortgage debt might prevent the utility from restructuring or splitting off operations at reasonable costs. We believe that the coming decade should be used to reduce outstanding mortgage debt so that the utility can restructure if need be, and to build up equity so that the utility can either function better in the competitive marketplace that may exist in the future, or build up the equity base that might be needed if utilities do have to join in a new round of construction.

Regulators who are still hung up on capitalization ratios of yore might consider using some hypothetical number for the utility operation but letting the utility do what it wants with the supposed surplus. If we want jobs and new business in this state, we will get them by encouraging some of the largest business entities in the state to keep the capital here, rather than exporting it to the rest of the world through decapitalization.

OPPORTUNITIES

Within two years, New York's utilities will end major construction efforts and enter a period of recuperation. The state can deal with this period in two ways. One response, which will be common throughout the country, will be to breathe a sigh of relief and go back to business as usual. The utility and regulatory structure will be left intact. Regulators will squeeze what they can out of the utilities. They will encourage the utilities to decapitalize themselves so as not to create equity ratios that upset the preconceptions of the regulators. The regulators will not worry about the next round of plant building or how it will be financed, because everyone knows that growth in demand will be low, nothing will need to be built, and even if that's not the case, somebody else will be on the commission when the problems resurface.

We view the coming decade as a period of opportunity rather than as a period of torpor. There's more to the power situation in New York State than how to distribute juice from the NYPA. Because of the financial and operating flexibility that is coming, there will be room to experiment and reorganize, to make the state a leader instead of prominently mediocre. Here are the opportunities:

- Create a regulatory framework in which market rewards and penalties are attached to incremental investment and in which the financial ratios to finance such investment become matters of market—not regulatory—judgment.
- Take advantage of low construction and high cash inflow to reduce reliance on mortgage debt and to create corporate structures that will be more flexible in a competitive marketplace.
- Look upon the existence of large, problem-free, cash-rich entities in the state not as problems for regulators but as growth centers that can add to New York's employment and business base.

A plethora of conferences and task forces are at work discussing and trying to solve problems that should have been tackled a decade ago. Water over the dam is just that, and no amount of recrimination or breast-beating will help to solve yesterday's crises. We should, instead, be looking toward the opportunities that are coming. *Carpe diem* is not the fish of the day at an Italian restaurant.

23

The Electric Utility Financial Crisis from a Banker's Perspective

Philip C. Kron

Citibank does business with the five largest electric utilities in New York State. As vice president of Citibank, I am familiar with the problems utilities in New York are facing, which are not much different from those being faced by utilities across the United States.

We are in a watershed period that has been caused by one overwhelming issue, namely, the nuclear plant construction problem—the real fallout from the accident at Three Mile Island. This issue is threatening the financial health of a broad spectrum of the utility industry. Fundamental viability is at stake in a number of instances, and at least two utilities, including the Long Island Lighting Company (LILCO) in New York State, have come dangerously close to filing for bankruptcy. In 1984, common stock dividends were reduced by at least six electric utilities in the U.S. and eliminated entirely by two others. In fact, these latter two, the Public Service Company of New Hampshire and LILCO, have taken the unprecedented step of eliminating all dividends on preferred stock as well. Even General Public Utilities (GPU), the owner of Three Mile Island, in its darkest hours, was never forced to consider this step. Bond and preferred stock ratings for many utilities have tumbled. Most utilities with nuclear construction programs are, at best, at the bottom of the investment grade category. A few are even classified as noninvestment grade. Some utilities can only finance in the "junk" bond market. Nothing tells the story of the financial deterioration of the electric utility industry better. Former AA and AAA securities are now being referred to as "junk"!

ELECTRIC UTILITY FINANCE BACKGROUND

The distinguishing financial feature of electric utilities is that they are probably the most capital-intensive companies around. This means they need large amounts of money annually to continue their essential service. Much of this money must come from external sources, namely, the capital markets. To put required external financing into perspective, capital expenditures in 1983 for the private investor-owned utilities amounted to $34 billion, with $16 billion, or 47 percent, raised externally. These are big numbers! No other industry has such an insatiable appetite for funds. And, just to give the facts a local dimension, New York electric utilities raised about $2.2 billion from external sources in 1983 and are expected to need $1.3 billion in both 1984 and 1985.

Part of the reason electric utilities need so much money every year has to do with the nature of the business. The other part relates to the accounting practices they are forced to follow. Regarding the accounting practices, historical depreciation and AFUDC—or "allowance for funds used during construction"—are the most onerous. They serve to keep rates and internal cash generation low. Correspondingly, they keep the need for external financing high. Historical depreciation is a problem because replacement costs far outstrip original cost. However, the really absurd accounting principle is the "pay later" convention of AFUDC. This accounting convention is used in lieu of setting higher rates today to cover the current financing costs of large construction projects. Essentially, what this means is that instead of getting current revenues to cover the debt and equity costs associated with building generating plants, such costs are capitalized. This means they are deferred and added to the cost of the project. They are then recovered through depreciation over the useful life of the asset once it is put into service. In other words, interest and dividend payments made today can only be recovered from ratepayers over the commercial life of the particular asset. This causes plants to be much more expensive when they finally come into rate base. It also gives rise to "rate shock." It places tremendous financial burdens on the particular utility building a plant and certainly detracts from the quality of reported earnings. A utility must not only finance the hardware costs of the plant because the depreciation allowances are so low, but it must fund the financing costs as well. For many utilities with large construction programs, AFUDC accounts for more than 100 percent of net income. This is certainly true for LILCO. AFUDC represented 109 percent of net income for common stock for the first six months of 1984. In fact, LILCO reported record earnings for the first six months of 1984. During this time it was on the very brink of bankruptcy. Somehow, record earnings and bankruptcy do not go together. This strange contradiction says more about the absurd AFUDC accounting convention than anything else I can think of.

THE ROLE OF BANK FINANCING

I would like to discuss the role of bank financing and how the financial problems of electric utilities are affecting us. Normally, banks provide short- and medium-term financing. This bridges electric utilities to periodic permanent financings in the public capital markets. Since electric utilities need to go to the capital markets on a regular basis, they usually build up short-term debt to a level where it can be refunded by a normal-size public offering. However, it is not unusual to use a layer of bank debt for a special purpose—like a coal conversion, or to help through a particularly heavy financing period. Also, bank debt is often used to get through periods of disruption in the capital markets, either with regard to price or availability. But banks are not long-term lenders, and therefore, since the payback on the investment in plants for electric utilities takes place over the long term, we must be assured that the capital markets are available. This means that anything that would disrupt access to the capital markets for other than a temporary period would be of major concern to us. This is why we are so concerned about the nuclear construction risk. We are concerned about the impact it may have on access to capital markets. Since banks traditionally fund themselves with short-term deposits, we cannot afford to find ourselves starting out as short-term lenders and find we have been converted to long-term investors.

THE NUCLEAR PLANT CONSTRUCTION PROBLEM

The nuclear plant construction problem is related to the issues of bankruptcy and access to capital markets. Readers may be familiar with the myriad of problems confronting those companies trying to complete nuclear plants. But, let me try to depict the issue in economic terms. Approximately 52 privately-owned electric utilities are involved with 41 nuclear plants still under construction. It has recently been estimated that about $80 billion has been invested in these plants. Approximately $40 billion is still to be spent. These are boxcar numbers! Many companies involved with these 41 nuclear plants have amounts in excess of their net worth tied up in these nonearning assets. LILCO, for example, has approximately $3.7 billion invested in Shoreham versus a common equity of about $2.3 billion. Obviously, the financial problems involved in not being able to complete these plants and to recover the investment in them are significant. These plants must be completed as soon as possible and brought into commercial operation.

But, what concerns me the most, as a commercial banker, is the possibility that the nuclear construction problem will cause the bankruptcy of an operating electric utility. Until recently, bankruptcy was an unheard of word or concept in the electric utility industry. However, as I mentioned earlier, the Public Service Company of New Hampshire and LILCO have come dangerously close to testing a process that has never been applied to a major operating electric utility. Even in the 1930s, it was the holding companies, by abusing the fundamental financial strength of the operating companies, that failed, not the operating companies themselves. Why does this concern me other than because it represents the possibility of having a large nonaccrual loan on my hands? I believe that the first bankruptcy of a major operating electric utility will destroy the most basic and underlying investment assumption associated with the electric utility industry, namely, that an electric utility will not be allowed to fail. It is this assumption that, in my opinion, has enabled the Public Service Company of New Hampshire to obtain commitments for over $400 million of financing. It has allowed LILCO and Consumers Power to raise $100 million or more each, even though the securities may be called "junk bonds." Large amounts of money are available to very weak companies because investors still believe they are "safe" investments and, by the way, because they carry very high coupons. Once the "no bankruptcy" assumption is destroyed, I doubt this kind of money, even at high rates, will continue to be available. The greater risk is that general capital markets may become inaccessible for most utilities trying to complete their nuclear plants. If this should happen, a broad spectrum of our electric utility industry could be brought to its financial knees. I must quickly point out that a number of my investment banking friends disagree with this doomsday scenario. They point to all the money that nuclear utilities have been able to raise despite an almost daily spate of bad news. However, my point is that we still have not had a bankruptcy. And, frankly, I'm not too anxious to test who is right. I think it is imperative that we keep capital flowing into the electric utility industry, and this strategy, in my opinion, means avoiding a bankruptcy.

It was this capital-markets spillover effect that caused the banks to work so hard to put a credit facility together in August 1983 to prevent LILCO's bankruptcy. Not only were we concerned about LILCO, but we were also concerned about the impact that LILCO's bankruptcy would have on other New York State utilities. Several have major funding requirements ahead of them relating to Nine Mile Point Two. I do not think there is any question that the bankruptcy of one utility in the state would have adverse effects on the state's other utilities. And, in fact, in one credit facility recently put together for a New York electric utility, a clause was included relating to the bankruptcy of another utility. Accelerated amortization of outstanding loans would be required if such bankruptcy had a material impact on our customers' access to capital markets. We are concerned about the spillover effect of the problems of one utility on other utilities. We are putting our customers and their regulators on notice to this effect. Commercial banks can only continue to finance electric utilities if we believe the long-term capital markets are open to those utilities.

After all the doom and gloom, let me hasten to back off a moment and give my assessment of where I think we are today regarding the nuclear construction problem, the problem that has brought the concept of bankruptcy to the electric utility industry. I think the bottom line is that the industry will survive. The lights will stay on. Construction and licensing problems are being overcome daily. Utilities are more on top of their projects and the projects are receiving more top management attention. Also, I believe the Nuclear Regulatory Commission (NRC) has adopted a more constructive approach toward resolving licensing problems and establishing steps necessary to gain approvals. Safety is not being compromised and regulations must be adhered to. But, I sense the NRC is working hard with many of its licensees to get the plants on-line. I think 1985 will prove to have been a critical year. It is the last major external funding year for the industry, and many plants are scheduled to be licensed. With these licenses successfully obtained, I think the industry will have dodged the nuclear bullet.

WHERE ARE WE GOING?

Some fundamental steps must still be taken to ensure that we will have the electric capacity we will need in the 1990s and beyond. First, foremost, and closest to home, the costs of the Shoreham nuclear plant must be dealt with in a way that will preserve the financial viability of LILCO. If this is not done, I believe that investment dollars will be severely curtailed for New York State's electric-generating capacity. This could very well have a negative impact on economic development within the state. Investors do not have to invest in New York State, or even in electric utilities generally for that matter, if the regulatory and political climate is not attractive. In order for LILCO's financial viability to be preserved, it must be able to recover substantially its Shoreham investment, whether the plant is able to operate or not. This means that the LILCO's customers must pay for the plant whether or not they are receiving any value from it. Therefore, assuming the plant is safe, the ratepayer would probably benefit by having it produce electricity and displace oil-fired power.

A less serious problem involves how those utilities with investments in Nine Mile Point Two recover the final cost of the plant. This will also affect the long-range attractiveness of investments in New York State utilities. There is already an "incentive rate of return" concept in place that penalizes stockholders for costs in excess of $4.6 billion—the latest cost estimate is near $5.5 billion. And the PSC instituted a "cap" in 1985 that would disallow any costs in excess of $5.4 billion. While I do not believe utilities should be given blank checks against their ratepayers, what benefit a stockholder gets out of new plant investment is unclear. If everything goes well, the best that can happen for stockholders is that their investments are merely diluted, because stocks must be sold below book value to build the plant (and all of the Nine Mile participants have stock trading below book value today). On the other hand, if things go badly and there are cost overruns or the plant never operates, the stockholders bear considerable downside risk. Sure, the ratepayer also bears some of the higher costs, but at least on the upside the ratepayer benefits from the provision of the electric power he or she needs or desires. These risk/reward issues are front and center in Albany relative to Nine Mile Point Two today. How they are resolved may affect the future adequacy of electric capacity in the state.

This then leads me to my third point. Some fundamental regulatory and accounting changes need to be made before it makes sense for an electric utility to embark upon any new major generation project, regardless of fuel type. The utility and its regulator must first agree that new capacity is needed. Then, the constructing utility must be assured of a current cash return on its investment in such generating capacity during the construction phase. In other words, "Construction Work in Progress (CWIP) in Rate Base" is mandatory for future projects. Those companies that pull through—and most, if not all, will—will not put their stockholders and ratepayers at risk again like they are today. What deferred accounting does is to inflate the cost of the plant to the point where the investment in a single plant equals or exceeds a given utility's equity. Also, deferred accounting subjects the utility's ratepayers to "rate shock" when the plant finally has to be paid for. AFUDC alone at the Shoreham plant amounts to $1.2 billion, almost a third of the total cost of the plant to date. And, worse, the cost of Shoreham is going up by more than $40 million a month because financial and operating costs that should be paid for currently are being deferred. The public cannot comprehend that $500 million a year of Shoreham's cost increases result from approved accounting practices. The current political and regulatory situation really acts as a major disincentive to any utility management to build a new plant of any type, much less for anyone to finance it. It's a "bet your company" proposition.

Finally, let me touch briefly on the nuclear option or, rather, the non-option. For a more complete discussion of this subject, I would refer you to a June 1984 Atomic Industrial Forum study entitled *"Nuclear Power in America's Future."*[1] I participated on the study committee. Among other things, we concluded that "nuclear power cannot at this time be considered a viable option on which to base new electric generating capacity in the U.S." until ways are "found to reduce the financial risks associated with capital-intensive and long lead-time construction." We concluded that "the private sector cannot take on the open-ended financial risks that now attend the nuclear power option." No new nuclear plants have been ordered since 1978 in the U.S. None will be ordered until the licensing process can be streamlined, construction periods halved, and financing can be accomplished on a prudent basis. This means that coal is our only viable central station generation option, despite what we read daily about problems with this fuel form. Frankly, I don't know how we are going to meet our future electricity needs. From a commercial banking standpoint, however, it is very doubtful that we would be willing to provide our normal bridge financing for any utility that wanted to start a new nuclear plant today.

CONCLUSION

I set out to describe financial constraints facing New York utilities. I hope I have accomplished this by discussing the financial dynamics of the utility industry and the importance of maintaining access to long-term capital markets. The only way commercial banks can continue to support the industry is by having confidence that utility securities will remain attractive investments to investors who have choices. Closer to home, this means that New York State utilities must remain financially viable so that the state can be assured of adequate supplies of power. Access—or lack of access, as the case may be—to capital markets is the ultimate financial constraint.

NOTE

1. Atomic Industrial Forum, *Nuclear Power in America's Future* (Bethesda, Md., June 1984).

24

Determinants of the Cost of Capital for the New York Power Authority —————

Ronald Forbes

The role of the New York Power Authority (NYPA) in New York's electric future is still evolving. With capital costs constituting a large portion of total electric bills for new generation and transmission facilities,[1] and with residential electric rates in New York already well above national averages, there will be a continuing incentive to expand the NYPA's role in bulk power supply. The NYPA's unique advantage has been its access to a lower cost of capital through the tax-exempt bond market. Maintaining ready access to this market on reasonable terms will be a key factor in defining the limits to the NYPA's future role.

The private markets where public debts are sold decide which borrowers will be accommodated from the limited pool of lendable funds. The terms offered the NYPA by the financial markets are strongly conditioned by the amount and types of competing claims offered investors by other borrowers, especially in the tax-exempt market sector. The terms offered the NYPA are also conditioned by the market's perception of the credit risks inherent in its projects.

This analysis focuses on the cost of capital to the NYPA and attempts to dissect the various elements embodied in the interest rates the NYPA pays on its debt issues. By way of summary and as a guide to the remaining analysis, the NYPA's cost of capital is determined by market factors, which relate to supply and demand conditions in the credit markets, especially the tax-exempt market; and "NYPA factors," which relate to the market's perception of the revenue stream from the NYPA's asset portfolio.

HISTORICAL OVERVIEW OF THE
COST OF CAPITAL TO THE NYPA

In 1954 the interest rate on the NYPA's first long-term bonds, with a scheduled maturity in 1995, was 3.200 percent. As the data in Table 24.1 show, the NYPA's long-term debt costs have more than tripled since then. In the October 1984 bond sale, the NYPA term bonds maturing in 2016 carried a yield of 10.425 percent. It is striking to note that 30 years ago, the interest rate on a 40-year debt was lower (by over 1.8 percent, or 180 basis points) than today's rates on a 30-day debt. The secular increase in borrowing costs for the NYPA carries an important message, namely, that the projects undertaken by the NYPA must promise a nominal return on assets that is also higher, and substantially so, than the return promised in 1954.

A fundamental point to be drawn from the data in Table 24.1 is that the cost of capital to the NYPA is driven by factors common to all capital markets and by those factors unique to the tax-exempt market. The Bond Buyer 20-Bond Index, a generally accepted index of long-term interest rates in the tax-exempt market, has also more than tripled since 1954. This index rose from 2.33 percent in 1954 to 11.71 percent in November 1981; it remained at a high level, 10.34 percent, as late as October 1984.

SUPPLY AND DEMAND IN THE
TAX-EXEMPT BOND MARKET

A perspective on the effects of supply and demand factors on tax-exempt interest rates is provided by the data in Table 24.2. The exemption to investors of interest income from federal income tax (and, for the NYPA, from New York State tax as well) confers special benefits in the form of lower interest rates on borrowers eligible to enter the tax-exempt market. The ratio of tax-exempt interest rates to taxable rates is one measure of the effectiveness of this market: lower ratios suggest more effective benefits to borrowers from the tax-exempt feature.[2] As indicated in Table 24.2, the tax-exempt/taxable yield ratio reached a low in the years 1977–79. The low relative yield ratio in these years was also translated into lower tax-exempt rates: the average annual new issue yield on AAA-rated bonds ranged from 5.20 to 5.92 percent during these years. It is also noteworthy that 1977 and 1978 were the only two years in the 1970s when the NYPA's term bond yield fell below 7 percent. One factor ex-

TABLE 24.1   The NYPA's Borrowing Costs, 1954–1984

NYPA Issue Date	Yield on NYPA Term Bond (Percent)	Maturity	Bond Buyer 20-Bond Index (BBI) (Percent)	Spread in Basis Points (NYPA–BBI)
Dec. 1954	3.200	1995	2.33	87
June 1956	3.000	1985	2.56	44
Jan. 1957	3.750	1985	3.23	52
Jan. 1959	4.200	2006	3.40	80
Jan. 1960	4.375	2006	3.78	60
June 1960	4.125	2006	3.52	61
Feb. 1961	3.750	2006	3.38	37
May 1964	3.550	2006	3.26	29
Nov. 1970	6.875	2010	6.28	60
Mar. 1971	5.625	2010	5.34	29
Sept. 1971	5.875	2010	5.39	49
Jan. 1972	5.500	2010	5.02	48
June 1972	5.500	2010	5.10	40
Sept. 1973	5.375	2010	5.34	4
Jan. 1975	7.875	2010	7.08	80
May 1975	8.125	2010	6.95	118
May 1976	7.500	2010	6.55	95
Sept. 1976	7.250	2010	6.52	73
Jan. 1977	6.625	2010	5.33	130
Apr. 1978	6.750	2012	5.69	106
Oct. 1979	8.000	2009	7.18	82
Sept. 1980	9.875	2020	9.18	70
Nov. 1981	12.125	2009	11.71	42
May 1983	9.750	2017	9.51	24
Oct. 1984	10.425	2016	10.34	9

plaining the relatively low tax-exempt interest rate during 1977-79 is supply. As noted in Table 24.2, net new issues of tax-exempt securities accounted for a relatively small proportion (less than 6.5 percent) of all credit market borrowing during this period. At the same time, institutional demand for tax-exempt securities from property and casualty insurance companies was at a peak. Casualty companies are the dominant institutional investors in long-term maturities, and demand from these institutions accounted for nearly half of net new tax-exempt securities issues in 1977-78.

TABLE 24.2 Selected Statistics on Demand, Supply, and Interest Rates in the Tax-Exempt Market

Year	Net New Tax Exempt Securities,[a] Total Market Share (Percent)	Percentages of New Tax-Exempt Securities Purchased by		Tax-Exempt Long-Term Interest Rate[c] (Percent)	Ratio of Tax-Exempt to Taxable Interest Rates[d] (Percent)
		Households[b]	Property/ Casualty Insurance Companies		
1971	10.3	1	20	5.22	71
1972	7.2	16	29	5.04	70
1973	5.9	36	24	4.99	67
1974	7.1	50	13	5.89	69
1975	7.3	39	16	6.42	73
1976	5.3	16	34	5.65	67
1977	5.7	1	49	5.20	65
1978	5.9	16	48	5.52	63
1979	6.1	39	33	5.92	62
1980	8.1	30	25	7.84	71
1981	6.6	58	18	10.67	73
1982	11.0	81	6	10.30	79
1983	8.9	91	−1	9.20	73

[a]Defined as net issues of tax-exempt securities as percentage of total net borrowing, all sectors. *Source:* Board of Governors, Federal Reserve System, *Flow of Funds.*

[b]"Households" includes mutual funds. *Source:* see note a.

[c]Average new issue reoffering yield on 20-year AAA rated General Obligation Bonds. *Source:* Public Securities Association, *Municipal Market Developments.*

[d]*Source:* Public Securities Association, *Municipal Market Developments.*

Since 1979, the yield ratio has increased significantly, rising from 62 percent in 1979 to 79 percent in 1982. The persistent shift in the tax-exempt/taxable yield ratio has signaled dramatic changes in both supply and demand, and has been accompanied by record high levels of tax-exempt interest rates.[3]

On the supply side, net new issues of tax-exempt securities have increased from 6 percent or less of all credit market borrowing to more than 8 percent and, in 1982, 11 percent of the credit market. These recent increases in the relative supply of tax-exempt securities are major factors explaining the relative increase in tax-exempt interest rates. Much of this new supply has been controversial; it represents uses of the tax-exempt market for the direct benefit of private enterprises or selected households. Table 24.3 points out, for example, that in 1983–84, industrial aid revenue bonds for private firms amounted to $22 billion or nearly 11 percent of all tax-exempt issues,[4] mortgage revenue bonds accounted for 17 percent, and nonprofit hospital bonds accounted for 7 percent of all issues. Many industrial development bonds are sold for the benefit of investor-owned utilities with investments in pollution control facilities, placing these utilities in direct competition with the NYPA for investors' funds.

Extending the benefits of tax-exempt financing to a broader constituency has diluted the benefits for all. Frankly, the public purposes served by some of these financings have been dubious and any derivative public benefits have been exceedingly indirect.[5]

In response to the burgeoning new supply of tax-exempt securities in recent years, the U.S. Department of the Treasury has proposed a major restructuring of the Internal Revenue Code to restrict severely the purposes and uses of future tax-exempt financing. These proposals, summarized in the Treasury Department's tax reform package,[6] would prohibit the issuance of new tax-exempt securities after 1985 if more than 1 percent of the proceeds were used directly or indirectly by any person other than a state or local government. If enacted, the ensuing reduction in supply would reduce tax-exempt interest rates relative to taxable rates; this in turn would reduce the cost of capital to the NYPA.

Other aspects of the Treasury Department's proposed reforms, however, may restrict the NYPA's flexibility in planning new capital financing programs. Certain proposals would require arbitrage earnings to be rebated to the federal government; they would also tie the timing of bond sales closely to the actual outlays of funds for construction; and they would prohibit advance refundings.

At the same time that relative supply increased in the last five years, the mix of investors was undergoing an historic change. Most notably, property and casualty insurance companies were withdrawing from the market. Commercial banks, faced with growing loan losses and rising costs from interest rate deregulation, also pulled back from new commit-

TABLE 24.3 Tax-Exempt New Issues Classified by Use of Proceeds (Volume in $Millions)

Use of Proceeds	1966–70 Volume	%	1971–76 Volume	%	1977–78 Volume	%	1979–80 Volume	%	1981–82 Volume	%	1983–84 Volume	%
Education												
Elementary and secondary	19,533	21.7	25,799	12.2	6,584	6.9	6,014	6.6	4,384	3.5	8,904	4.3
Higher education and other	7,066	7.8	12,077	5.7	3,443	3.6	3,718	4.1	4,744	3.8	8,693	4.2
Highway transportation	10,033	11.1	12,447	5.9	2,896	3.1	2,214	2.4	2,638	2.1	3,314	1.6
Water and Sewer	9,680	10.7	18,057	8.5	6,571	6.9	6,090	6.6	7,879	6.3	12,497	6.0
Electric, gas, other utilities	3,909	4.3	25,714	12.2	9,695	10.2	9,211	10.0	15,943	12.8	19,117	9.2
Housing	1,639	1.8	13,794	6.5	9,391	9.9	27,605	30.1	20,270	16.2	35,859	17.3
Hospitals and health facilities	3,108	3.4	15,038	7.1	8,212	8.7	7,458	8.1	14,878	11.9	15,250	7.3
Parks, civic centers	1,763	2.0	5,372	2.5	2,401	2.5	2,341	2.6	1,946	1.6	1,743	0.8
Ports, airports, other transportation	6,958	7.7	9,500	4.5	3,532	3.7	2,830	3.1	7,051	5.6	8,865	4.3
Industrial aid (including pollution control)	4,387	4.9	13,600	6.4	9,036	9.5	7,784	8.5	15,400	12.3	21,957	10.6
Refunding, advance refunding	N/A		N/A		21,234	22.4	3,693	4.0	5,456	4.4	26,124	12.6
Multipurpose and all other	22,037	24.6	60,537	28.6	11,902	12.5	12,745	13.9	24,431	19.5	45,438	21.9
TOTAL	90,113	100	211,395	100	94,896	100	91,703	100	125,020	100	207,761	100

Source: Data files maintained by the Municipal Finance Study Group, State University of New York at Albany.
Note: N/A, Not applicable.

ments in the so-called bank range maturities (up to ten years). As a result, the growing supply of new tax-exempt securities has by necessity been marketed to individual investors and their surrogates—mutual funds and unit trusts.

Institutional demand is not likely to revive in the near future. Property and casualty insurance company purchases of tax-exempt securities are closely correlated with profits from insurance underwriting; as underwriting profits increase, insurance companies have more funds available to invest. These profits peaked in 1978; since then insurance underwriting losses have increased each year from $21 million in 1979 to more than $4 billion in 1981, $11 billion in 1983, and over $12 billion in 1984.

Marketing to individuals requires significant adjustments in interest rates and in the costs of new issues. Individual investors are, at the margin, in lower marginal tax brackets than are institutional investors. Tax-exempt yields must rise therefore to provide a yield equivalent to the after-tax yield on taxable investment alternatives for these low-bracket investors. Marketing to individual investors also imposes higher information costs on issuers and their investment bankers. Smaller transaction sizes and the costs of search for investors result in higher underwriting costs.

In addition, individuals are less likely to have the time or expertise to anaylze lengthy and complex security provisions. They rely instead on bond ratings and news media for guides to credit quality, and they monitor secondary market performance as a guide to marketability and liquidity. In contrast to a marketplace dominated by large institutional investors, a market that relies on large-scale distribution networks for small investors needs different approaches. The NYPA, for example, would probably benefit from a formal investor-relations program designed specifically to attract and retain the support of the individual investor.

THE "NEW YORK EFFECT" AND THE NYPA'S COST OF CAPITAL

Supply and demand factors are at work within the sectors of the tax-exempt market as well. These intramarket forces often result in significant differences in interest rates among tax-exempt securities that vary by the region or state in which a borrower is located. The reasons for these local market effects are several. State laws, for example, generally foster a demand for local government tax-exempt securities through tax provisions that exempt within-state issues from that state's income taxes; income earned on holdings of out-of-state issues, on the other hand, is taxed. Moreover, state banking laws often require collateral on public deposits to be in the form of within-state securities. For many investors, the preference for local issues is based on the expense required to monitor changing credit conditions. These information costs generally increase as the geographical distance between the investor and the borrower increases.

The data in Table 24.1 demonstrate that the yield on the NYPA bonds in 1975 was 118 basis points above the national tax-exempt market average as measured by the Bond Buyer Index. Since reaching its high in 1977, this yield spread has steadily been reduced, reaching a low of 9 basis points in October 1984. The dramatic narrowing of the NYPA's cost of capital relative to other tax-exempt securities is largely the result of changing supply and demand conditions for all New York tax-exempt securities. In 1975, the entire market for tax-exempt securities from New York was thrown into disarray because of serious payment difficulties associated with New York City and the State Urban Development Corporation. Many interrelated factors combined to create the climate for these financial emergencies in 1975, including a severe national recession and its impact on a state economy that had long lagged behind the nation's growth. Among the important antecedents of the crisis, however, was the mounting use of public debt by governmental agencies in the state. As the data in Table 24.4 indicate, total new issues from New York amounted to nearly 20 percent of all tax-exempt bonds sold nationally in 1975. During the early 1970s the state, many of its statutory authorities, and certain of its local governments borrowed heavily to finance complex and controversial projects, to balance unsynchronized cash flows, and to cover budget deficits. In an apt summary of these practices, the Moreland Act Commission in 1976 noted:

> The present credit crisis in New York State has arisen in large part because the public debt of the State, its agencies and authorities has been allowed by its leaders to grow more rapidly than the State's ability to support the debt. While the State's economy was slowing down during the past decade, these leaders were permitting debt accumulation to speed up instead of putting on the brakes. Optimism clashed with reality and responsibility was the victim. Even the financial community ignored this undermining of the State's credit until zero hour.[7]

One consequence of the questionable quality of this rapid new supply can be noted in risk premiums required to sell full faith and credit general obligations of the state: as noted in Table 24.4, the consequence of this state risk program was a lofty 111 basis point yield spread in 1975. Although the NYPA's debt obligations were (and are) legally distinct from those of the state and its local governments, the general malaise associ-

**TABLE 24.4 Selected Statistics on
Tax-Exempt Securities in New York State**

Year	Volume of New Bond Issues from New York[a] ($Billions)	Percentage of All New Tax-Exempt Bond Sales	Yield Spread (Basis Points): State General Obligation Bonds versus the Bond Buyer Index[b]
1975	5.8	19.7	111
1976	4.0	11.8	70
1977	5.8	12.9	29
1978	3.9	8.4	24
1979	2.9	7.0	-10
1980	3.5	6.9	49
1981	2.9	6.2	-27
1982	5.3	6.8	-35
1983	5.2	6.2	-49
1984 (June)	2.7	8.1	-55

[a]*Source:* Public Securities Association.

[b]Calculated as the average annual spread between the yield on 20-year maturities of state new issues and the Bond Buyer 20-Bond Index.

ated with any New York name carried over: the NYPA risk premiums jumped in 1975 and remained high through 1978.[8]

Since 1975, there has been a remarkable turnaround in debt management practices in New York. The supply of new issues declined from $5.8 billion in 1975 to $2.9 billion annually in 1979 and 1981, and New York's share of national tax-exempt new issues had dropped to 6.2 percent by 1981.

In contrast to the recent slowdown in the volume of new issues, there has been a significant increase in the demand for New York tax-exempt income since 1975. One measure of this demand is the growth in personal income taxed at high marginal rates. Between 1975 and 1981, the aggregate reported income for individuals living in New York and filing federal tax returns with adjusted gross incomes of about $100,000 expanded from $4.0 billion to $13.9 billion, which represents a compound rate of 23 percent a year. Gross income of individuals living in New York and filing federal tax returns with incomes between $50,000 and $100,000 grew from $5.5 billion to $19.3 billion—also an annual rate of 23 percent—over the same period.

The juxtaposition of declining supply and increasing demand has been translated into a steady reduction in yields on New York tax-exempt securities relative to the market. As Table 24.4 shows, the yield spread on state general obligation bonds dropped from a positive spread of 111 basis points in 1975 to a negative spread of 55 points in 1984. Table 24.5 points out that the changing fortunes of the New York label have encompassed other borrowers as well. Bonds of the Municipal Assistance Corporation for the City of New York (MAC), for example, carried an average premium of 240 basis points above the Bond Buyer Index from 1975 to 1977; this spread declined to 130 basis points from 1978 to 1981, 78 points in 1982, and only 7 basis points in 1984.

In summary, the improving market for the NYPA bonds over the last decade can be largely ascribed to a vastly improved climate for public debt within the state of New York. The lessons to be drawn from this review are that the cost of capital to the NYPA will continue to be strongly conditioned by overall credit market factors and by local market forces. The NYPA does face a separate set of financial constraints, however, and these constraints are important in maintaining access to markets on terms that are favorable under prevailing conditions.

FINANCIAL CONSTRAINTS ON THE NYPA'S FUTURE ROLE

An operational definition of a financial constraint that serves as an indicator of market access for individual borrowers, and one that summarizes many of the factors that determine credit quality, is the debt service coverage ratio. This ratio is defined as:

$$\text{Coverage Ratio} = \frac{(\text{NOR} + \text{DEP})}{(\text{INT} + \text{PRIN})}$$

where NOR = net operating income, DEP = depreciation, INT = interest expenses, and PRIN = annual bond principal payments.[9] As defined, the coverage ratio imposes a cash flow constraint on publicly sponsored utilities that otherwise might pursue cost minimization as a key objective.[10] Moreover, the close relationship between bond ratings and coverage ratios emphasizes the practical significance of this cash flow constraint. As noted by Standard and Poor in its discussion of key financial indicators on tax-exempt revenue bond issuers: "The key factor is debt coverage. We have always felt that the magnitude of debt service coverage was not as important as the stability and predictability of such coverage."[11]

Table 24.6 provides some data on selected governmental utilities that support this link between cash flow and bond ratings. For the five utili-

TABLE 24.5 Selected Tax-Exempt Bond Issues from New York State

Sale Date	Issuer	Term Maturity	Yield (Percent)	Rating by Moody/ Standard & Poor	Bond Buyer 20-Bond Index (Percent)	Yield Spread (Basis Points)
9/30/82	Triborough Bridge & Tunnel (Convention Center)	2012	10.875	A	10.48	40
10/15/82	MAC	2008	9.700	Baa-1/A	9.25	45
10/14/82	MTA (Transit)	2012	10.400	Baa/BBB+	9.25	115
12/20/82	MTA (Service Contract)	2013	11.000	A/A−	9.84	116
12/8/82	Port Authority, NY-NJ	2017	10.180	A1	10.13	5
1/14/83	New York City	2003	11.400	Ba1/BBB	9.37	203
2/18/83	Metropolitan Transportion Authority MTA	2003	10.250	A/A−	9.53	72
		2013	10.500	A/A−	9.53	97
5/24/83	NYPA	2003	9.500	A1/A+	9.51	−1
		2017	9.750	A1/A+	9.51	24
6/15/83	NYS Medical Care Facilities Agency (Suffolk Hospital)	2016	10.500	Baa/BBB+	9.38	112
6/6/83	Dormitory Authority (Columbia)	2013	10.180	AA/AA−	9.69	49
5/9/83	MTA Service Contract	2003	9.150	A/A−	8.86	29
		2015	9.400	A/A−	8.86	54
2/9/83	MAC	2008	10.875	Baa1/A	9.74	114

Date	Issuer	Year	Price	Rating	Yield	
7/20/83	New York City	2013	10.300	Ba-1/BBB	9.54	76
9/1/83	MAC	2003	9.875	Baa1/A	9.75	13
		2008	9.950	Baa1/A	9.75	20
10/14/83	Triborough Bridge & Tunnel Authority	2003	9.625	Aa/A+	9.67	-4
		2012	9.950	Aa/A+	9.67	8
10/7/83	SONYMA	2013	9.500	AA/–	9.49	1
1/19/84	MAC	2008	9.700	A/A	9.60	10
2/2/84	MTA	2004	9.625	A/A–	9.51	12
		2017	9.875	A/A–	9.51	37
4/18/84	Dormitory Authority (Society of New York Hospital)	2003	9.675	A1/A	9.89	-22
		2015	9.800	A1/A	9.89	-9
4/27/84	MAC	2008	10.000	A/A	9.94	6
5/4/84	Dormitory Authority (U. of Rochester)	2003	9.875	A1/A+	9.99	-12
		2009	10.000	A1/A+	9.99	1
5/9/84	Dormitory Authority (Cornell)	2004	9.900	AA/AA	10.19	-29
		2014	9.970	AA/AA	10.19	-22
7/5/84	SONYMA	2004	10.750	AA/–	10.69	6
		2011	10.875	AA/–	10.69	19
7/19/84	NYSERDA (Niagara Mohawk)	2014	11.250	Baa1/A–	10.36	89
7/24/84	New York City	2014	10.950	Baa/BBB	10.19	76
8/2/84	MAC	2008	9.875	A/A	9.92	-5

TABLE 24.5 (*Continued*)

Sale Date	Issuer	Term Maturity	Yield (Percent)	Rating by Moody/ Standard & Poor	Bond Buyer 20-Bond Index (Percent)	Yield Spread (Basis Points)
8/17/84	SONYMA	2004	10.000	Aa/–	10.02	–2
		2009	10.180	Aa/–	10.02	16
8/30/84	MTA Transit Facility	2004	10.100	Baa/BBB+	10.17	63
		2014	10.930	Baa/BBB+	10.17	76
9/26/84	MAC	2008	10.320	A/A	10.15	17
9/21/84	MTA Service Contract	2004	10.080	A/A–	9.95	13
		2014	10.250	A/A–	9.95	30
10/4/84	NYPA	2005	10.250	A1/A+	10.34	–9
		2016	10.430	A1/A+	10.34	9
10/19/84	MTA Transit Facility	2001	11.000	Baa/BBB+	10.24	76
		2011	11.375	Baa/BBB+	10.24	114

Source: Public Securities Association.
Legend: MAC: Municipal Assistance Corporation for the City of New York.
MTA: Metropolitan Transit Authority.
SONYMA: State of New York Mortgage Agency.
NYSERDA: New York State Energy Research and Development Authority.

ties with AA bond ratings, the average debt-service coverage ratio attained in fiscal year 1983 was 2.02. By contrast, the seven utilities with single A ratings carried an average coverage ratio of 1.47 times debt-service.

The data in Table 24.6 add a final perspective on the importance of the coverage ratio and bond ratings. The cost of capital to any borrower carries a premium to reflect perceived credit risks. Interest rates, therefore, tend to be significantly correlated with bond ratings. An analysis of public power bonds issued in 1983 and 1984 documents this association. Bonds with at least one AA (or better) rating carried interest rates that were, on average, 31 basis points higher than the Bond Buyer 20-Bond Index. Bonds with single A ratings from both rating agencies carried yields that averaged 74 basis points above the index; while noninsured bonds with at least one rating below A carried an average yield premium of 154 basis points above the Bond Buyer Index.

Table 24.7 provides a recent history of the coverage ratio on the NYPA projects. These data indicate that the overall financial results place the NYPA below the AA bond rating category. Viewing the NYPA as an integrated utility, overall debt service coverage ratio has ranged from 1.25 to 1.54, which is characteristic of utility bonds with single A ratings. The NYPA's overall coverage ratio, therefore, does serve as a constraint at present. Future project selection as well as the management of existing projects must be conducted within a constraint that does not allow significant reductions in future cash flow coverage.

It is useful to analyze further the elements that produce the overall coverage ratio and to assess their impact on the NYPA's cost of capital. Over time, the NYPA has implemented significant changes in its asset portfolio mix and in the structure of its long-term indebtedness.

Students of finance have long been conditioned to the concepts of risk and return. It is well understood, for example, that layering significant financial leverage—that is, the proportion of debt in the capital structure—on assets whose returns reflect a high degree of operating risk can magnify the volatility of net cash flows. At the same time, finance theory has refined its concepts of asset management to incorporate the role of the benefits (and limits) of portfolio diversification in reducing the volatility of returns on asset portfolios.

The NYPA's asset portfolio has become more diversified over time, but this diversification has increased—not decreased—the volatility of cash flow coverage. The NYPA's first capital projects, the so-called 1954 resolution projects, consisted of the Niagara and St. Lawrence hydroelectric generating facilities. The second major projects included the Fitzpatrick nuclear project and the Blenheim-Gilboa pumped storage power project. This latter group of projects was originally financed under a separate bonding program referred to as the 1970 Resolution Bonds. In 1974,

TABLE 24.6 Selected Features on Selected Public Power Bonds

Issuer	System type	Standard & Poor's Rating	Historical Debt Service Coverage	Anticipated Capital/Bonding Needs	Other Key Features
Los Angeles	Integrated	AA –	1983: 1.82×	$1.6 billion	Strong additional bonds tests; offset by large volume of off-balance-sheet financing; strong service area and historical financial performance.
Salt River Project	Integrated	AA	1982: 2.02× 1983: 1.92× 1984: 1.85×		Interest during construction financed from current revenues; strong financial performance; diversified resource mix; sound economic base.
Sacramento Municipal Utility District (SMUD)	Integrated	AA	1981: 1.64× 1982: 2.01× 1983: 2.90×	$1.5 billion	Strong financial performance and strong service area; future project portfolio is diversified.
Nebraska Public Power District: Electric System Revenue Bonds	Integrated	A	1981: 1.19× 1982: 1.23× 1983: 1.21×	$245 million to 1987	Low anticipated debt needs; stable financial performance.
South Carolina Public Service Authority Electro System Bonds	Integrated	A +	1984: 1.83×	$900 million to 1992	Heavy future debt burden; lack of diversification in customer base; favorable fuel mix.
Jacksonville Electric Authority	Integrated	AA	1983: 1.60×	$1.8 billion	Strong historical financial performance; strong cash position; sound economic base.

Power Authority of the State of New York General Purpose Bonds	Wholesale	A+	1983: 1.34× 1982: 1.54× 1981: 1.39×	$618 bonding to 1989; $250 million possible entitlement purchase	Sound financial performance; favorable resource mix. "The authority has been involved in discussions regarding possible involvement in Nine Mile Point 2 and Shoreham."
Omaha Public	Integrated	AA–	1981: 1.40× 1982: 1.41× 1983: 1.84×	$210 million	Low anticipated debt needs; stable financial performance.
Lincoln, Nebraska, Electric Distribution System	Integrated	A+	1981: 1.38× 1982: 1.41× 1983: 1.34×		Low anticipated debt needs; sound financial operations; strong service area.
Grand River Dam	Wholesale supply	A–	1981: 1.08× 1982: 1.14× 1983: 1.11×		Significant construction risk; business risk; weak historical financial performance.
Lower Colorado River Authority Priority Revenue Bonds	Integrated, primarily wholesale	A	1982: 1.65× 1983: 1.55×	$2 billion to 1993	Wholesale customers' contracts (85% of revenues) expire in 1999.
Austin Texas Combined Utility System Prior Lien Bonds	Integrated, also water and sewer	A+	1981: 1.85× 1982: 1.73× 1983: 1.88×	$3 billion to 1990	Significant construction risks in nuclear project; strong historical financial performance and economic growth.

Source: Standard & Poor, *Credit Comment: Utility Revenue Criteria Update,* October 15, 1984. "Key Features" represent a summary of comments by Standard & Poor's analysts.

NYPA undertook a third major capital program, purchasing generating units from consolidated Edison (ConEd). This purchase was financed by a third bonding program referred to as the General Purpose Bond resolution. An indication of the incremental volatility is noted by observing the fluctuations in debt coverage for the 1970 resolution projects. As noted in Table 24.7, debt service coverage ratio on the 1970 projects fluctuated from 0.47 in 1977 to 1.15 in 1978 and 0.88 in 1979.[12]

At the same time that its asset portfolio risk has been increasing, NYPA has been sending mixed signals to investors regarding the structural ties between portfolio cash flow and the level and sources of payments to bondholders. The rate covenant, for example, is a commonly used contractual agreement between enterprises and their bondholders; its purpose is to constrain the propensity of public officials to minimize charges to constituents for services. In the bond resolution for the 1954 projects, the rate covenant pledged the NYPA to the levy of rates sufficient to produce a net cash flow at least equal to 1.4 times debt service. The resulting surpluses benefitted bondholders in two ways: the substantial margin over required principal and interest outlays eased investors' concerns over credit risk, and the application of excess funds to accelerate debt retirement improved the secondary market for NYPA bondholders. For projects financed under the 1970 Bond Resolution, the rate covenant was reduced to an effective debt service coverage requirement of 1.25. The 1974 General Purpose Resolution further reduced the rate covenant to 1.15.

Under normal circumstances, attempts to lower the contractual commitment to generate revenues, especially on projects that carry high oper-

TABLE 24.7 Debt Service Coverage Ratio for the NYPA

Year	Combined Projects	Coverage Ratio 1954 Projects	1970 Resolution	General Purpose Resolution
1977	1.25	2.48	0.47	1.15
1978	1.44	2.68	1.15	1.15
1979	1.35	2.60	0.88	1.16
1980	1.51	3.44		
1981	1.39			
1982	1.54			
1983	1.34			

Source: Moody's Municipal Credit Reports, based on various financial statements presented on cash basis.

Notes: Coverage ratio on 1954 projects based on mandatory principal maturities. Following 1979, financial results were combined for projects financed under the 1970 resolution bonds. Financial results are presented on a combined basis only.

ating risks and high financial leverage, would dramatically lower bond ratings and increase the cost of capital. However, at the same time that the NYPA was reducing its rate covenant, it unified, from the perspective of bondholders, its asset portfolio cash flow. The 1970 and 1974 resolutions developed a workable form of "system" financing that pledged the revenue capacity of the overall NYPA asset portfolio to meet debt service obligations. Under these resolutions, excess cash flows from prior projects (e.g., the 1954 project) were pledged (subject to limits) sequentially to expand the resources available to meet debt service on more recent projects. Figure 24.1 diagrams these contractual arrangements as they existed in 1979. A more unified financing developed as a result of

FIGURE 24.1 Flow of Funds from Authority Projects after Retirement of 1954 Bonds

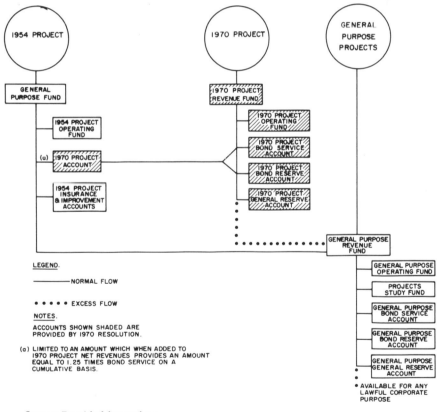

Source: Provided by author.

the retirement of the 1954 bonds and the 1980 refunding that defeased[13] the outstanding 1970 Resolution Bonds. These debt management operations effectively tied together the financial structure of cash flows from the NYPA's asset portfolio to its bondholders. This system-financing structure is now perceived by many as the essential component of the NYPA's market access because it permits shortfalls in revenues from one sector of the asset portfolio to be offset by the excess revenue capacity from other assets. This portfolio effort can reduce the volatility of cash flows to bond holders. Asset/liability management under the systems approach takes on added importance for the NYPA because, unlike many if not most wholesale power producers, the NYPA does not have long-term power sales contracts with distribution systems.[14]

Today the NYPA has a well-developed asset/liability management structure that can facilitate access to markets in the future. Important hurdles, however, remain to be overcome. There are outstanding legal challenges to the system-financing structure that are retarding the full potential of the system-financing approach. Until these challenges are resolved, the NYPA's future investment opportunities and its market standing will be somewhat constrained.

NOTES

1. A recent study concluded that the average increase in first-year required revenues for utilities with nuclear construction work in progress was 35 percent. See Aaron S. Gurwitz and Daniel E. Chall, "Nuclear Power Plant Construction: Paying the Bill," Federal Reserve Bank of New York, Quarterly Review (Summer 1984):55.

2. At the federal level, this ratio is at the heart of an ongoing policy debate. Briefly, the debate focuses on the loss of federal revenues because of the tax-exempt status of interest paid on some bonds. At high levels of the tax-exempt rate and at high levels of the tax-exempt/taxable yield ratio, the loss of revenues is magnified. For a summary of the issues in this debate, see David Beek, "Rethinking Tax-Exempt Financing for State and Local Governments," Federal Reserve Bank of New York, Quarterly Review (Autumn 1982):30–40.

3. As an example of the effect that changes in the relative yield ratio have on the level of the tax-exempt rate, applying the 1979 yield ratio of 62.00 percent to 1982 results in a tax-exempt rate of 8.08 percent, which is more than 200 basis points below the actual rate in 1982.

4. Many small issue industrial development bonds are placed privately with local commercial banks and are not publicly reported. This "underground" market may more than double the publicly reported volume reported in Table 24.3. See Department of the Treasury, Office of Tax Analysis, "Treasury Report on Private Purpose Tax-Exempt Bond Activity during Calendar Year 1983," March 28, 1984.

5. For an overview of these trends in tax-exempt financing, see Ronald Forbes, Philip Fischer, and John Petersen, "Recent Trends in Municipal Revenue Bond Financing," in Efficiency in the Municipal Bond Market, ed. George Kaufman (Greenwich, CT: JAI Press, 1982).

6. See *Tax Reform for Fairness, Simplicity and Economic Growth*, Treasury Department Report to the President, vol. 2, November 1984.

7. *Restoring Credit and Confidence*, Report to the Governor by the NYS Moreland Act Commission on the Urban Development Corporation and Other State Financing Agencies, 1976.

8. In the perceptions of some, the decision to finance the purchase of Astoria and Indian Point Three did associate the NYPA with the (mis)fortunes of New York City.

9. This ratio is appropriate for utilities on an accrual basis of accounting. For utilities that prepare financial statements on a cash basis, the appropriate cash flow is net operating revenue.

10. This definition—cost minimization subject to a debt-service coverage constraint—may be contrasted with investor-owned utility objectives, which can be expressed as profit maximization subject to a regulated rate of return constraint. For a model of public utilities that follows the approach in this paper, see Darwin Hall and Brian Thomas, "A Financial Model for Publicly-Owned Electric Utilities," *Energy* 9, no. 4 (1984):333–40.

11. Standard & Poor, *Ratings Guide*, pp. 271–72.

12. Following general practices, the NYPA sells additional bonds to fund a debt service reserve fund that can be and has been (1977, 1979) used to meet temporary shortfalls in cash flow.

13. The term *defease* refers to the practice whereby an escrow fund is established with monies sufficient to meet fully the debt service on outstanding obligations.

14. Long-term power sales contracts are designed to shift the risks of great volatility to distribution systems that must pay for power regardless of whether that power is used (take or pay contracts).

PART VII

INSTITUTIONAL AND REGULATORY CONSTRAINTS —

The authors in this final section have each served, and one is still serving, as commissioners responsible for the state or federal regulation of electric utilities. After reciting the too-long litany of current regulatory problems, each of the authors moves on to survey an array of proposed changes and improvements, many of which would alter fundamentally the way economic regulation is conducted and, in some cases, would encourage more competition in the electric supply industry. All three deplore reforms, primarily legislative, to move regulatory commission structures and procedures further toward the judicial model. Such changes are seen as at best shortsighted and at worst as crippling to the commissions' ability to perform the tasks for which they were originally established, namely, to provide economic regulation of a natural monopoly in the public interest. It is precisely because the technology of electricity supply is complex and continually changing, these authors emphasize, that their regulatory oversight bodies must have the freedom and power to adapt and modify their regulatory procedures accordingly, and judicial bodies simply are not designed to initiate or synthesize. Because strict judicial bodies are merely receptors that make reactive decisions about the equity claims of others, they are poorly suited to accomplish effective economic regulation.

All three authors also acknowledge that the public mistrusts regulators and electric utilities, but Kahn and Schuler both emphasize that this is an inevitable consequence of regulation and should provide an incentive to get rid of unnecessary regulation. Kahn also points out that during particular periods in the past when the public was first less and then more outraged about regulators and the electric utilities, those sentiments in both cases were probably a result less of the performance of those institutions and more of external economic, demographic, technological, and social forces. Nevertheless, all three authors acknowledge that many changes could be made that would improve the likelihood of a more desirable outcome from the regulatory process.

Both Schuler's and Kahn's chapters offer a wide scope of prescriptions for regulatory change; Schuler also raises the additional problem of integrating and coordinating economic and environmental regulation. Stalon explores the impact of the increasingly formalized, judicial nature of the regulatory process; in the course of his discussion he alludes to the inane consequences (except to the municipal electric companies who receive "preference" hydro-power) of the legalistically motivated rulings of the Federal Energy Regulatory Commission in assigning hydro-power rights. Kahn joins in disparaging the economic consequences of those rulings, which may allocate more Niagara and St. Lawrence hydro-power to those few customers who are already the primary beneficiaries of that low-cost power. Stalon emphasizes that those regulatory results are an inevitable outcome of quasi-judicial procedures and that moves to codify the process further would only make matters worse. Finally, both Kahn and Schuler acknowledge that it has been the cost-plus nature of setting regulated utility prices that diminishes utility incentives to cut their costs, and they advocate forms of price-comes-first regulation to help reinstitute incentives for efficient operation of these regulated utilities.

25

The Institutional and Regulatory Structure for Providing Electric Service: A Conceptual Basis for Change

Richard E. Schuler

INTRODUCTION

As a regulator, I was always perplexed by the intensity of public outrage over most electric rate increases during the 1970s. Obviously, nobody likes rising prices, but in the past era of widespread inflation when all energy sectors were particularly hard hit, why weren't there riots at supermarkets and gasoline stations like those that erupted at utility rate hearings? Of course part of the answer lies in our increased dependence on this product. Another source of the problem is that electric service is provided by a private monopoly, albeit a regulated one. Perhaps rising prices are more palatable when consumers have alternatives and when all choices are exhibiting similar price trends.

In retrospect, however, I feel the greatest source of the problem is what I have come to call the myth of regulation, namely, the idea that somehow regulators have the mystical power to hold costs down. In fact, they *can* hold prices down, at least in the short run. And so it would be quite easy to be a popular regulator. The fear, of course, is that with inadequate funds, utility service would go the way of New York City subways, and in the long run the public would be left with neither adequate service nor low costs. In fact it is extremely difficult for regulators to control the underlying cost of doing business; that is dictated by much broader market pressures. This is one reason why I loudly applaud all efforts to deregulate the prices charged by a particular industry if it is no longer in a position to harm the public. The extent to which consumers are deluded by the myth of regulation may in fact do much more public harm than the exertion of any limited vestigial market power by the industry.

Unfortunately, the recent moderation in the escalation of oil and gas prices will not totally eliminate the future problems of electric utility regulation, because the overall inflation of the past decade has inserted a delayed time bomb into the electric utility pricing process. While most public attention has been focused on the rapid escalation of the construction costs of nuclear power plants, fossil fuel-fired facilities have also experienced dramatic price increases over the past decade. Figure 25.1 shows that up until 1970, while overall economy-wide construction costs were rising more rapidly than the general consumer price index (CPI), utility power plant construction costs were increasing less rapidly than those economy-wide costs. Over the past decade, this relationship has been completely reversed, and construction costs of fossil fuel-fired plants have escalated more rapidly than economy-wide cost indices. Furthermore, the recent diminished growth in demand for electricity will not eliminate these inflationary impacts because as older plants wear out, they must be replaced with far more expensive new units. So, even without any further inflation in fuel prices or power plant construction costs, electric rates will continue to increase over the next 30 years, even for utilities experiencing zero load growth, as old $100/kW power plants are replaced by new $1,200/kW units. Thus most utility regulatory bodies will be the bearers of bad tidings for many years to come.

REGULATORY ANOMALIES

How to deal with the inflated costs of all new generating plants is the general overriding problem, then, that must be dealt with in the regulatory arena during the coming years; however, there are a number of other anomalous results of the existing regulatory structure that also illustrate the need for improvement or modification of the existing structure. The following five examples emphasize specific current regulatory problems:

1. The "money-saving" rate increase.
2. Oswego Five and Six: more expensive electricity and also more pollution.
3. Pollution swaps in New York City.
4. The planning horizon versus the demand horizon.
5. Backdoor natural gas deregulation through PURPA.

1. Customers in New York State Electric and Gas Corporation's service territory experienced a substantial "money-saving" rate increase when the 650 MW Somerset coal-fired plant came on-line in 1984. Somerset is not currently needed because of shortages in capacity; nevertheless, customers are told by the utility that over its lifespan, the plant will

FIGURE 25.1 Construction Cost Escalation, Fossil Fuel-Fired Electric Generating Plants

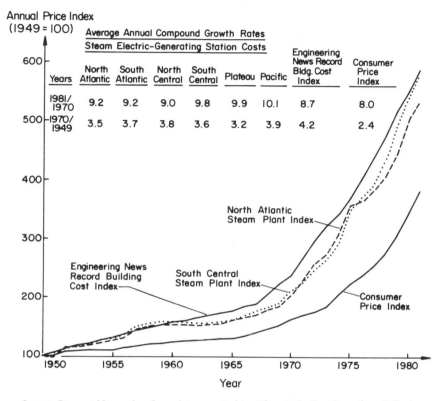

Years	Average Annual Compound Growth Rates Steam Electric-Generating Station Costs						Engineering News Record Bldg. Cost Index	Consumer Price Index
	North Atlantic	South Atlantic	North Central	South Central	Plateau	Pacific		
1981/ 1970	9.2	9.2	9.0	9.8	9.9	10.1	8.7	8.0
1970/ 1949	3.5	3.7	3.8	3.6	3.2	3.9	4.2	2.4

Source: Computd by author from data reported in, "Quarterly Cost Roundups," *Engineering News Record*, (March, June, Annuals for years 1950–1980).

save them money because it burns fuel more efficiently and uses less-costly fuel sources than the other alternatives. Question: If this plant, which is generally considered to have been efficiently constructed, will save customers money over its useful lifespan, why can't it save them money today? Answer: It might, if utility accounting and rate-making procedures were to undergo vast changes. (Note that customers served by Shoreham and Nine Mile Point Two will also experience these money-saving rate increases, if and when those facilities ever go on-line.)

2. The operating practices at Niagara Mohawk's (NIMO) Oswego oil-fired station on Lake Ontario over the past several years, through a combination of environmental laws and economic regulatory practices, resulted in both more pollution and higher electric costs than may have

been necessary. Yet, given the constraining regulations, NIMO no doubt was trying to operate in a way that would result in the lowest possible expense to customers. This is a case where two adjacent units, one completed before the Federal Clean Air Act went into effect, the other after, had substantially different upper limits on the sulphur content of the fuel oil they were permitted to burn. Oswego Five, the older unit, could burn relatively high sulphur-content oil; Oswego Six, the newer unit, was required to burn extremely expensive low sulphur-content oil. Attempting to minimize the cost of electricity generated, the utility operated Oswego Five heavily and Oswego Six rarely. The net result was both more pollution and higher-priced electricity than was necessary had the utility been able to burn fuel with a slightly higher sulphur content in Oswego Six in exchange for using lower sulphur-content fuel in Oswego Five. Nevertheless, most environmental interests, justifiably concerned about recent retrenchments in environmental programs, strongly resisted attempts to lower the permissible standards at Oswego Six.

3. New York City has been a leader in pushing for tough environmental regulation, and it implemented even more stringent sulphur-emission standards than mandated by the Federal Clean Air Act. As a consequence, Consolidated Edison (ConEd) was forced to convert its coal-fired generating units into oil-fired facilities in order to achieve the desired low level of emissions. Then, faced with the 1974 oil embargo, ConEd's fuel prices soared, as did the price of electricity. Therefore, diesel-powered co-generation facilities became economically viable for a large number of ConEd's customers. The net environmental tradeoff is a diminution of high-level sulphur oxide emissions in exchange for increased low-level carbon monoxide emissions. In this case, who is better off: the New York City pedestrian or the boater off Cape Cod?

4. In the 1960s many utilities experienced severe shortages because electricity demand was doubling every seven years, a growth rate far in excess of the previous decade. Nevertheless, because most new plants could be completed within 3–5 years, virtually all capacity shortages were eliminated by the early 1970s. However, in that decade demand *growth* fell to 0–2 percent for most of New York State's utilities, while the total planning and construction time for completing a typical power plant increased from a 4–5 year range to 10–12 years. Question: If demand patterns can change more rapidly than the time it takes to build a power plant, does this guarantee sequential episodes of chronic over- and undercapacity?

5. Recently an enterprising private developer announced plans to construct a 200 MW gas-fired generating facility near Syracuse. The plant would sell waste heat in the form of steam to Syracuse University, thereby qualifying as a co-generation facility under the Public Utility

Regulatory Policy Act of 1978 (PURPA) and enabling the developer to sell all generated electricity to Niagara Mohawk at that utility's avoided costs. This entrepreneur's ingenuity is to be applauded; however, note that the net effect is to circumvent the current federal regulation of the wellhead price of old gas. Through this scheme, the developer is essentially using PURPA to deregulate natural gas, since if that gas were sold into interstate pipelines for ultimate purchase by gas distribution companies, the wellhead price would be regulated. Question: Is that gas more valuable in the hands of gas distribution companies or in the hands of Niagara Mohawk in the form of electricity?

For readers who may have thought that there were few conceptual problems remaining to be solved in the electric utility regulatory arena— that all we need are some hard-nosed regulators to apply a few simple principles consistently—it is hoped these illustrations emphasize the frequently complex, unintended consequences of simple regulations. In the remainder of this chapter, several basic principles of regulation are reviewed to serve as a basis for examining particular proposals that might be used to deal with these anomalies.

BACK TO BASICS

In constructing a path for the future, it is frequently useful to think about how and why a particular industry ought to be regulated, if it were possible to start from scratch. In the case of electric utility regulation, there are four primary reasons for regulation. The first is the original basis: that decreasing-cost industries will naturally evolve into monopolies. While as a sole supplier a monopoly can provide the least expensive service, its potential to gouge the public through overpricing that service warrants intervention by the public sector. In the early days of the industry, it was difficult to argue that a second reason for public intervention in the provision of electric service was because reliable service was a public good. Today, however, this is far easier to justify; particularly in large urban areas, since as we have seen from the two major blackouts in New York City over the past twenty years, unreliable service has the potential of generating a collective public harm that is far greater than the individual private losses. The third rationale for public regulatory intervention is the negative externalities—air, water, and potential radiation damage plus noise—that may extend geographically far beyond the immediate neighborhood of the power-generating facilities. The final rationale for government intervention in electric supply involves the idea of equity, and again this is a recent concern. In the early days of the industry, the service was not a necessity. But with the increased urbanization of America, which has been closely linked to and made possible by increased electrification (e.g., subways, lighting, elevators, air conditioning), some use of electricity has become essential for all Americans. A fundamental administrative tension has arisen as a result of this nationwide dependence. Since regulatory bodies normally charged with economic regulation are not accustomed to gauging equity issues, those reallocative decisions have traditionally been left to legislators. However, recognition of the necessity of some minimum quantity of electricity for virtually every household has thrust these equity issues on many regulatory bodies, particularly in the form of proposals to establish "lifeline" rates.

Historically, however, the dominant reason for regulating electric utilities has been their decreasing-cost characteristics. Certainly the distribution of electricity through overhead lines or underground conduits along public thoroughfares in urban areas has exhibited this characteristic. One line can usually serve all the neighbors in a block just as effectively as two, and at half the cost. Furthermore if overhead facilities are the dominant source of supply, the visual blight associated with multiple lines crisscrossing the streets is not likely to win applause from anyone but the most avid technocrat. Nevertheless, while few would argue for having multiple local suppliers of electricity in concentrated urban areas, there is little evidence that strong scale economies exist over widespread geographic regions.[1] Thus, considering the distribution function alone, there appears to be no reason why any individual organization needs to serve populations in excess of 250,000–500,000 people in order to approximate the lowest possible cost.

On the other hand, improvements in the generation of electricity have been the dominant source of scale economies over the first 100 years of the industry, which is primarily why the industry evolved from many small localized utilities to the large integrated corporations of today. However, even adjusting for inflation, if one looks at the history of construction and operating costs for new large generation facilities built over the past 10–15 years, one can seriously question whether additional scale economies in generation are achievable. This intuition is confirmed by an econometric study by Laurits Christensen and William Greene of system-wide generation costs, which suggests that all additional scale economies in generation are realized by a system whose peak demand is as small as 3,000 MW.[2] This is the current size of the Long Island Lighting Company (LILCO), and it is substantially smaller than Niagara Mohawk's or ConEd's peak demands. The fact that no individual generating unit larger than 1,200 MW has been proposed in the past decade, in light of the previous steady escalation in unit sizes, provides additional technical evidence that the possibilities for further scale economies in generation may have been exhausted.

Substantial scale economies that still exist are in the transmission, the long-distance bulk power transportation function, of electricity. There appear to be few practical limits to increasing the voltage of transmission lines substantially above existing levels, and higher voltages permit far greater quantities of power to be dispatched at modest increases in costs. Furthermore, since the object is to transport power while maintaining a high degree of system reliability, duplicate facilities must be made available in the event of forced outages or planned maintenance. But with the enormous capacity of a 765 kilovolt (kV) line, parallel paths may be built most economically on a statewide basis, and in the case of the Northeast, perhaps through a multistate planning program. Thus, if we were starting over again with economic regulation of utilities, it would certainly make sense to plan transmission systems on at least a statewide basis; the distribution entities, on the other hand, might be structured as smaller, communitywide systems.

What about the generation of electricity? With a statewide peak demand in excess of 20,000 MW, it takes approximately twenty of the largest, most efficient possible units to serve New York State's needs. Given the interconnection between each of these generating units and all of the state's distribution customers through a statewide transmission grid that functions somewhat like an interstate highway system, why couldn't each generating unit be represented by a separate independent, unregulated company? Certainly twenty generation companies represent a far larger number of competitors than we currently have providing automobiles or most major consumer durable products like appliances. In fact, power is currently dispatched by the New York Power Pool (NYPP) as if such a competitive system existed, so operational difficulties should not be a major objection. The only real questions that need to be answered are: would anyone be willing to risk the capital to build a new large generating facility in a deregulated bulk power market; and would any, significant magnitude of what Joskow and Schmalensee have called "economies of integration" be lost if generation and electric distribution companies were not one and the same entity?

Surely no one questions that some governmental authority is needed to establish the standards and enforce the rules regarding maximum tolerable levels of pollution. This is a proper governmental function, and the only open issues are: What level of government should decide these issues? Should economic and environmental regulation be integrated? What should be the method of implementation and enforcement?

Which level of government should decide hinges in part on the geographic extent of the spillover. As an example, particulate emissions do the greatest damage to nearby neighbors, whereas sulphur dioxide is an immediate threat as well as a substantial contributor to a more widespread geographic insult in the form of acid rain. Carbon dioxide and greenhouse effects, furthermore, are truly a global concern. These differential impacts suggest that different effluent regulations should be assigned to various levels of government, depending upon the nature and extent of the insult. To some extent this is the current regulatory case.

By comparison, some of the anomalous results indicated in the early part of this chapter could be resolved if economic price and environmental regulation were integrated on a day-to-day basis. However, very different constituencies have grown up to support these two different types of regulatory bodies, so that today a great deal of mistrust exists between the two. The sensible resolution of economic and environmental electric supply dilemmas has become more of a political than a regulatory problem.

The third question, concerning the form of implementation, was settled in the early days of environmental regulation when legal prescriptions regarding emission standards and means of control were enacted into law. Nevertheless, there is an increasing sense that these legal prescriptions frequently lead to silly results, so economists' proposals for a system of properly structured effluent charges are once again receiving modest attention.

APPLICATION OF MARKET PRINCIPLES TO REGULATION

While the previous conceptual analyses suggest that at least the generation of electric service may be amenable to the efficiency-enhancing incentives of market pressures, incorporating these pressures into the regulatory framework and/or actually implementing limited degrees of deregulation raise difficult thorny issues.

Procedural Differences between Regulation and Market Processes

Under regulatory pricing procedures, the commission normally sanctions construction of a project, the utility builds it, and then the regulatory body reviews the costs to see if they were prudently incurred. This immediately places the regulatory body in a post facto mode of operation. The utility knows that unless it has made egregious errors of judgment and management, most of the costs incurred will be approved, and therefore its investors will have a fair shot at receiving their anticipated rates of return through the subsequent prices set by the regulatory body. Thus to utility investors, the security of future payments has been considered almost as certain as investing in public-sector infrastructure projects such as water supply, roads, and sewerage systems. In contrast, in a competitive market situation market pressures set the price, and the potential investor must estimate whether a competitive facility can be built at a sufficiently low cost to afford a fair return at those market-determined prices; the sequence is completely reversed. *Under regulation cost determines the price; in a competitive market the price, to some extent, determines the cost.* In the long run, of course, even in competitive markets costs establish the upper bound on price. Any firm's pricing position is always restricted by the costs of the next higher priced firm. If firm A were to let prices drift above the costs of firm B's production, firm B could undercut firm A, stealing away its customers.

Nevertheless, it is this price-comes-first ordering of competitive markets that creates the incentives for firms to find new and more efficient ways to produce their services. If regulatory bodies are to provide similar incentives to encourage production efficiency, they too must find mechanisms to establish prices long before the full, exact cost structure is revealed.

Proposals to Mirror a Market under Regulation

One group of attempted changes, while hardly qualifying as fundamental alterations in the institutions that regulate and provide electric service, nevertheless does represent significant changes in regulatory procedures. These changes include a set of incentive modifications that have already been implemented by the New York State Public Service Commission (PSC). One modification is the incentive rate of return (IROR) attached to the 1982 opinion authorizing five of the state's utilities to continue construction of the Nine Mile Point Two nuclear power plant under the condition that a target completion cost be established.[4] Under the terms of the IROR, the utilities would have to swallow 20 percent of any cost overruns, but they could retain 20 percent of any cost savings. The second example is a 1983 decision to modify the operation of NIMO's fuel adjustment clause,[5] so instead of compensating the utility directly and month by month for its actual incurred fuel expenditures, these costs would be projected one year into the future. Subsequently, the utility would have to absorb a percentage of any actual fuel cost that was higher than the estimate, but it could retain a percentage of any cost savings. Both of these initiatives are modest steps toward establishing a price-comes-first rate-making procedure, and both provide the utility with incentives to alter its method of operation to produce the service at an actual cost that is lower than the original guidelines.

Another frequently proposed alteration of existing regulatory accounting practices involves implementing some form of economic depreciation. One perceived benefit of these alternative rate-setting schemes is to avoid the "money saving" rate increase outlined earlier. Recently these alternative depreciation schedules have been proposed as a means of phasing in large nuclear plants that have experienced substantial cost overruns. In either case, the primary motive is to avoid the severe "rate shock" that has been frequently felt in recent years as new large generating facilities are brought on-line and placed into the rate base. The fundamental conceptual argument against straight line depreciation, which is customarily used in setting utility rates in part because of accounting expediency, is that straight line depreciation establishes utility price patterns over time that are similar to those that might arise under a market process, *only by sheer coincidence*. Figures 25.2, 25.3, and 25.4 illustrate the long-term price patterns that emerge if utility rates are established using straight line depreciation. Several alternative growth scenarios are represented.

Advocates of economic depreciation argue that in a competitive market, a developer who builds an apartment project, anticipating that rents will rise over the next 20 years with inflation (depending on the magnitude of the initial construction cost of the project), has no guarantee that rents in the first several years will cover the capital cost of the project. In fact, many speculative real estate ventures show negative cash flows in the early years of operation. The market sets the allowable rental charges; however, if the developer is correct, as those rents rise over time substantial profits will be earned in the future. Advocates of basing utility prices on some form of economic depreciation argue similarly that if a nuclear plant is built to save energy costs in the future, then it simply

FIGURE 25.2   Conventional Rate Treatment for a Single Plant

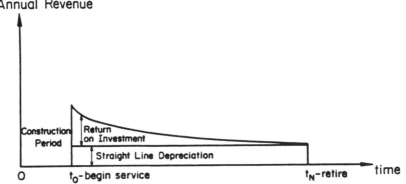

Source: R.E. Schuler, "Alternative Electric Power Plant Financing and Cost Recovery Methods: Introduction, "*Resources and Energy* 7, no. 1 (March 1985):3.

**FIGURE 25.3 Conventional Rate Treatment for
Utility with Growing Demand: A New Plant Each Year**

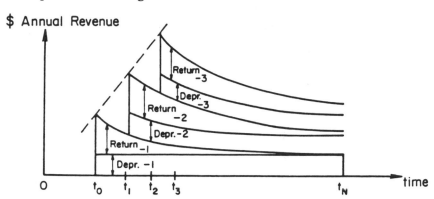

Source: Schuler, "Alternative Electric Power Plant Financing."

does not make sense to charge customers prices in the present that are far greater than the cost of generation by alternative available sources. By using economic depreciation, it is claimed, the revenues to the utility would be escalated gradually over time as operating savings materialized; and coincidentally, the investors' returns would accelerate over time. As summarized in a recent special edition of *Resources and Energy*,[6]

**FIGURE 25.4 Conventional Rate Treatment for Utility with
Constant Demand: Five-year Staggered Plant Replacement**

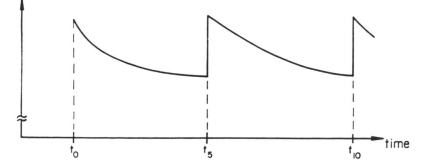

Source: Schuler, "Alternative Electric Power Plant Financing."

the salient remaining questions associated with implementing economic depreciation include: What are the actual computational mechanisms used to establish the price structure over time? What risk premium, if any, might investors demand on the cost of capital as a result of deferred cash recovery?

The first recent examples of competitive provision of generation are the projects developed under the 1978 PURPA. According to this act, potential developers of small hydroelectric, co-generation, or other small generation facilities based on renewable energy resources are assured of a market for the electricity they produce; the nearest utility must purchase their power at the utility's avoided cost. For independent developers, this closely approximates selling in a competitive market, since the highest price that any firm can hope to charge in a competitive market is bounded by the most costly alternative. This is exactly the definition of avoided cost that has been used in New York State and many other jurisdictions in establishing buy-back rates for these small independent power project developers. Furthermore, the powerful incentives of the price-comes-first procedure are also operable, since the independent developer must estimate future avoidable costs and then determine whether or not a competitive facility can be constructed. The lower the cost of construction, the higher the profits to the small project developer.

William Baumol has also proposed an alternative price-comes-first rate-making procedure for utilities.[7] Under Baumol's scheme, the utility and the regulator would be subject to traditional rate hearings occasionally, but they would also select a representative utility cost index that neither party had the power to manipulate. Rates would then be allowed to escalate or decline automatically between formal hearings in accordance with the index, up until some unspecified time in the future. A cost-cutting incentive arises because, having some predetermined range of revenues for the foreseeable future, the utility can implement cost-cutting techniques with some assurance that it will be able to retain those savings for some period of time. In the long run, the utility's customers will also benefit because when there is a new formal rate proceeding, the PSC will adjust rates to take into account the observed cost-cutting moves. In order to preclude strategic gaming between the parties, the future date when a new formal case will be initiated must be selected randomly. Accordingly, the New York State lottery might hold a drawing one day each year to see which utility would be subject to a formal rate proceeding.

The interesting aspect of all of these proposals is that they edge ever closer to initiating a competitive market for bulk power supplies. In fact, by extending PURPA guidelines to *all* new electric generation facilities, not just those built by small independent developers using renewable energy sources, an effective competitive market could be established for the future provision of bulk power supplies. What is needed, of course, is a spot market for power generation *capacity* sales that is comparable to the existing NYPP transactions for energy *flow* sales. Under such a situation, avoided costs would become the pricing standard for the state. All utilities and private developers would look at the current avoided costs and make their best projections into the future; then, if they felt they could construct new facilities and beat those costs, they would. Alternatively, if no one felt able to match the current avoided costs, no new construction would emerge, and capacity prices in spot market transactions would rise until some enterprising firms were willing to take a gamble and begin construction. Under such a scheme, if conservation or load management were the more economical means of adjusting capacity, those obviously would be the solutions selected.

Of course what has just been described is nothing more than a competitive market for bulk power transactions, and because of the oligopolistic nature of these firms, some residual market power would be retained (although certainly no more than that which exists in the automobile or appliance industries today). In a separate analysis Benjamin Hobbs and Richard Schuler have estimated that for upstate New York, in the short run, deregulation might lead to price increases of anywhere from 2 to 10 percent over existing marginal costs.[8] However in the long run, as more facilities were erected in response to these market stimuli, those price premiums would be reduced to 1-2 percent. These calculations make no attempt to estimate what production efficiency improvements might arise as a result of the incentive spurs of competition; they merely attempt to estimate the upper price risk of deregulation.

However, in order to hold the short-term overall price increases to within the 2-10 percent range, an additional mechanism must be established in the transitional period, keeping utilities that already own power plants that were built at low construction costs many years ago from receiving unwarranted windfall profits. One way of resolving this problem is to retain some form of interim regulation over the price at which a utility is permitted to sell to itself its own generation from an existing plant that was constructed under the previously existing regulatory scenario. Nevertheless, such a scheme does seem to be operationally feasible, and it provides all the needed incentives to minimize future cost increases.

Recent Industry and Regulatory Restructuring Proposals in New York State

The utility industry's Empire State Power Resources, Inc. (ESPRI), proposal of the mid 1970s, which would have established a statewide consortium of private utilities to aid in the financing of new large generating facilities, appears to be a move diametrically opposite to the likely consequences of the "mirror-the-market" conceptual proposals of the preceding section. While the New York State PSC turned down the ES-PRI proposal on jurisdictional grounds (subsequent regulation of wholesale power transactions between intrastate utilities would have been preempted by the Federal Energy Regulatory Commission, [FERC]), there may have been a sounder economic justification for making that decision if competitive pressures are to be the wave of the future.

Ironically, most of the recent legislative proposals to alter electric utility regulation in New York State, like similar plans around the country, are really attempts to tinker with the PSC's procedures and/or to restructure its organization, but such proposals do not deal with the substantive issues of whether and to what extent we should regulate, or what form regulation ought to take in the future. As an example, the proposal by New York State Senator Martin Auer to regroup the regulatory functions of a number of state agencies can be viewed as a form of horizontal reorganization. In his proposal, the PSC's communications division would be combined with the New York State Cable Commission under one telecommunications regulatory body. Furthermore, the electricity and gas regulatory functions of the PSC would be combined with the State Energy Office (SEO) to establish an energy regulatory commission. While there is a certain functional logic to this proposal, its weakness is in not acknowledging the substantial cross-fertilization that spills over from experience in regulating one industry into practices within another. Thus lessons learned in dealing with "cream skimming" by competitive carriers in telecommunications should prove invaluable in establishing practices for gas pricing (where oil is the competitor) and for electricity as more competitive pressures emerge in the future. While the technology may vary widely between these very different regulated industries, the conceptual regulatory practices should be similar, and a test of the consistent application of these principles to different technologies is a very useful check on the regulator.

A second recent proposal is Assemblyman Oliver Koppell's plan to separate the existing functions of the PSC into three separate entities: a quasi-judicial body, the commission; a consumer-oriented trial staff; and an enforcement agency. While this proposal may have legalistic merit, it is predicated primarily on a judicial view of the commission's functions. Sensible regulation requires the flexibility to pursue a regulatory body's legislative powers in order to alter and adapt its procedures continually as the nature of an industry's costs and its consumers' demand for service change over time. In my view, the inherent weakness of this second plan is that it strips the commission of the internal adaptability that is essential if the PSC is to modify policies applicable to an evolving industry.

The other major regulatory initiative that has arisen at the national level is an attempt to modify the Public Utility Holding Company Act of 1935 (PUHCA). In my view, many of these proposals would take a giant step backward by allowing electric utilities to escape effective state regulation by forming interstate holding companies. While billed as a necessary prerequisite to allow utilities to diversify into nonregulated businesses, the desirability of that diversification might vary from state to state, and the decisions should not be preempted by the federal government, as might be the result under several recent proposals. Furthermore, if utilities are so interested in escaping the long arm of state regulators, why don't they forcefully advocate the deregulation of bulk power sales? Then no commission would have a reasonable basis for questioning the diversification of unregulated generating companies.

However, none of these proposals seems to contemplate market-like incentives. By far the largest problem confronting any deregulated attempt to build a large, central station generating plant is the long time lag that exists between identifying the need for a new facility and its completion. Few private investors would be willing to put up money for 10–12 years before they could anticipate any cash returns, which is the case under current regulatory practices. Nevertheless, much of that elapsed time is not devoted to the actual construction of facilities but, rather, is associated with the planning for and acquisition of the necessary environmental permits associated with site selection and approval. In no way do I mean to suggest that an attempt should be made to short-cut these careful environmental reviews. What does need to be investigated is the possibility that the time duration when private investment is at risk can be reduced while those proper environmental evaluations are undertaken.

INTEGRATING ENVIRONMENTAL EVALUATION
WITH ECONOMIC REGULATION

Of course, identifying environmentally acceptable generation sites is very much a decision about what constitutes the public interest and as such needs to be carefully thought out and thoroughly debated. Since the quality of the environment is a shared attribute, decisions about it and/or mechanisms to achieve environmental objectives should ultimately be decided in the public arena. However, the most severe problem confronting any scheme to deregulate the generation of electricity has to do with whether or not entrepreneurs would be willing to assume the enormous costs and risks that are currently associated with the environmental planning for new generating plants. In fact it is the assumption of those planning risks by the current regulated utilities that in many cases has caused their severe financial distress. The solution does not lie in short-cutting those evaluations. Nor does it lie in accommodating the demands of many utilities to be fed cash continually during construction of a plant in order to withstand all unforeseen planning exigencies. The proper solution may arise by placing those decisions and risks about public goods where they properly belong: in the public sector. One possibility is to have the state, through an exhaustive hearing process, identify and acquire suitable sites throughout the state, each targeted for a particular type and capacity of electrical generation that is compatible with the local ecology and resource supplies. The idea for such a "site bank" is not new, but it has lain dormant since there have been few recent applications to build new power plants.

Were a backlog of sites identified and acquired by the state, subsequent commercial and residential development in the region could be undertaken with full knowledge of the future possible construction of nearby power plants. In acquiring these sites, the state could compensate existing neighbors for their loss, if any. This site identification and land acquisition process would be costly to the responsible state agency; nevertheless, a portion of those costs might be recouped by auctioning off the sites to subsequent power plant developers. The premise supporting government site assembly in this instance is no different than the arguments made in favor of government land acquisition for urban renewal projects. And while the overall site costs might be no less expensive for private generation developers than they are under current procedures, the risk would be greatly reduced since there might be fewer unanticipated impediments to construction. Not only would the establishment of this backlog of sites reduce the risk to private power plant developers by reducing the elapsed planning and construction time of facilities to 3–5 years for a coal-fired unit, it would reestablish a planning horizon more compatible with the demand horizon described at the beginning of this chapter.

The remaining environmental anomalies laid out in the beginning of this chapter can be reduced through careful, comprehensive environmental regulation. As an example, an undesirable substitution of carbon monoxide for sulphur dioxide should occur only in those situations where the environmental oversight placed on low-level emissions from stationary diesel generators is not as stringent as regulations pertaining to sulphur dioxide emissions by utilities. The bigger problem, curing the Oswego Five and Six dilemma, would be exacerbated under a system of economic deregulation of bulk power supply prices. Under present regulations the costs of operating existing generation units, most of which were completed prior to the effective date of the Clean Air Act and were therefore exempted from its provisions, are lower than the costs of operating new facilities that must incorporate pollution control devices or burn more expensive, lower-sulphur-content fuel. As a consequence, a powerful incentive exists to retain existing facilities and not to build new, less polluting replacements. One way of restoring a balance between environmental regulation and the economically efficient production of power is to extend the provisions of the Clean Air Act so that they cover all generating facilities. Another method for resolving that problem is to revert to the economist's prescription for dealing with environmental insults: apply effluent charges to the emissions from *all* facilities. With a similar level of effluent charges applied to both Oswego Five and Six, NIMO would operate both units in such a way as to minimize combined private economic and public environmental costs. And under deregulation, each utility's avoided costs would then reflect the environmental costs of generating with existing facilities, either without pollution control devices but paying high effluent fees, or with the expense of retrofitted devices, whichever is less costly. Developers would weigh those proper avoided costs against the cost of constructing new facilities with abatement devices. But, sensible economic regulation of utilities can only arise if there is sensible environmental regulation as well. The two must move hand in hand, not on separate tugging and pulling courses.

SUMMARY: REGULATION AS A SYSTEM OF INCENTIVES

What I have provided is a conceptual framework for gauging proposed alternative regulations as well as a series of practical examples to test that framework. The discussion emphasizes that electric utility regulation must continually be thought of as a system of incentives. Every major new regulatory thrust seems to yield unintended consequences, many of which are worse than the original problem that led to the new prescription. Worst of all, in many instances the practice of regulation encourages the publicly held myth that something good can actually be accomplished. Therefore eager candidates frequently stumble over each other trying to make things better, but they actually make matters worse through the perverse incentives that frequently emerge from a maze of regulation. Where the fundamental need for regulation has diminished over time, perhaps the only sensible regulatory approach is to reverse the historic policy direction and snip the regulatory cord.

What role can existing private utilities play under the scenarios outlined above? The same that they played in the past, if they have the talent and motivation to construct new plants at lower costs than their competitors. What role would exist for the New York Power Authority (NYPA)? That is one potential provider of the services that comes under the "public function" heading in the previous discussion. At a minimum its responsibility should be to ensure the evolution and successful operation of the interstate highway system of bulk power supply, the transmission grid. It might also make sense for the NYPA to develop a bank of generating sites for its own use or to be auctioned off to other private or regulated developers. And finally, under a deregulated scenario, should large nuclear plants or other enormously capital-intensive, evolving technologies prove to be the wave of the future, then given the financial and social risks associated with constructing those large facilities, it might make sense to have some public agency be the demonstrator of their economic viability. Surely private, profit-seeking providers will be increasingly reluctant to take the multibillion dollar gamble of attempting to complete large nuclear power plants if less risky alternatives are available.

Finally, none of these regulatory evolution schemes can operate without the close cooperation of federal regulatory bodies with individual state agencies. That does not imply that all future economic and environmental regulation of electric utilities should be determined by the federal government; indeed what has been presented is a plea for sensible, flexible regulation based upon basic principles rather than rigid rules. This means that it might make sense if the form of regulation varied geographically in accordance with underlying demographic, economic, environmental, and technological conditions. By allowing the states to take the lead in setting regulations, the case for diversity will be established, and then the federal government will only need to intervene where overriding principles and national concerns are at stake.

NOTES

1. See the study by Leonard Weiss, "Antitrust in the Electric Power Market," in *Promoting Competition in Regulated Markets*, ed. A. Phillips (Washington, D.C.: Brookings Institution, 1975).

2. L. R. Christensen and W. H. Greene, "Economies of Scale in U.S. Electric Power Generation," *Journal of Political Economy* 84 (1976):655–76.

3. P. L. Joskow and R. Schmalensee, *Markets for Power: An Analysis of Electric Utility Deregulation* (Cambridge, MA: MIT Press, 1983).

4. See New York State Public Service Commission, "Opinion and Order Concluding Inquiry into Financial and Economic Cost Implications of Constructing the Nine Mile Point No. 2 Nuclear Station," Opinion 82-7, Case 28059 (Albany, NY, April 16, 1982).

5. See New York State Public Service Commission, "Opinion and Order Concluding Proceeding on Motion of the Commission as to the Rates, Charges and Rules and Regula-

tions of Niagara Mohawk Power Corporation for Electric Service (Phase II–Fuel Adjustment Clause)," Opinion 83–17, Case 27741 (Albany, NY, September 19, 1983).

6. R. E. Schuler, ed., "Alternative Electric Power Plant Financing and Cost Recovery Methods," *Resources and Energy* 7, no. 1 (March 1985).

7. W. J. Baumol, "Productivity, Incentive Clauses and Rate Adjustment for Inflation," *Public Utilities Fortnightly* 110, no. 2 (July 22, 1983):11–18.

8. B. F. Hobbs and R. E. Schuler, "Deregulating Electric Utility Generation: Estimates of Price and Welfare under Spatial Oligopoly," Cornell University working paper #325 (September 1984).

26

Analysis and Synthesis in Quasi-Judicial Multimember Regulatory Agencies————————

Charles G. Stalon

"You don't lobby FERC, you lawyer FERC."[1]

A regulatory commission cannot rise much above its staff, and a commission staff cannot rise much above its commissioners. That proposition deserves special attention as commissions find themselves regulating firms that are increasingly competitive. Those interested in understanding the quality of decision making of regulatory commissions must understand the peculiar symbiotic relationship between staff and commissioners. I use the word *peculiar* to emphasize both the distinctive nature of this relationship and its odd and eccentric nature. I believe the following observations are relevant to certain other government agencies such as the Nuclear Regulatory Commission (NRC), but my focus is on the traditional economic regulatory agencies.

My thesis can be stated simply: I believe that there are sound reasons to conclude that the current organization, composition, and structure of today's energy regulatory agencies make many of them incapable of performing a pervasive leadership role.[2] I argue that such agencies all too often lack the ability to synthesize information well enough even to oversee their own operation; they lack the tools, especially with respect to electricity, to offer real leadership at the national level; and these bodies simply by their nature lack the attitudinal commitment (guts, if you will) to make realistic forecasts. The major reasons for this failure are the peculiar relationships among commissioners and the relationships between commissioners and the commission staff. Both of these relationships are substantially determined by law.

One principle that is intuitively appealing is that the internal organization of every government agency should be determined primarily by the purposes of the agency. This principle is less helpful in understanding multimember quasi-judicial regulatory agencies than it first appears to be, for two reasons.

First, the purposes of the agency are seldom explicitly stated in enabling legislation, so one of the responsibilities that most regulatory agencies must accept is to define and articulate explicit purposes and to build an organization around these definitions. Few commissions do this explicitly, and doing it implicitly is a long process, one that is seldom completed for most commissions whose members change frequently.[3]

At this point it is useful to identify my views about the principal purposes of an economic regulatory agency. Fundamentally, an economic regulatory agency's objective is to bring about economically efficient prices for regulated goods and/or services. Within that overall objective, it must understand and implement sound economic theory, it must maintain a familiarity with and sensitivity to the relevant regulated and nonregulated firms and markets so as to be able to apply that theory, and its decisions must be final and timely so as to allow sellers and buyers to make decisions necessary to their economic activity. Moreover, it should stimulate competitive situations where competition is constructive and limit competition where it is likely to be destructive. The last obviously requires that the commission understand the difference between the two. At times regulators are given other specific tasks, such as licensing and allocating valuable resources by nonmarket criteria. In such cases, the commission should execute the tasks in a manner that minimizes the inefficient use of society's resources.

Regulatory agencies are also often charged with functions of coordination and cooperation with other governmental agencies. The multimember agency has particular problems in accomplishing this sort of work, the most acute of which is the inability of a single member to bind the others. Another less obvious problem is that cooperation means that the parties will act in harmony. However, regulatory commissions, for reasons I will explain shortly, cannot predict their own decisions. This inability to coordinate efforts with those of other agencies in closely related fields is a serious bar to a commission's ability to play a leadership role, even within the government.

Second, regulatory organizations are not only required to regulate in pursuit of ill-defined objectives, they are required to collect information, to deliberate, and to make decisions in conformity with politically approved methods. At the federal level and in most large states, these procedural rules are very restrictive. With regard to the effects of the administrative procedure acts (APAs) and open meetings acts (OMAs) on multimember, quasi-judicial bodies, three consequences deserve mention.

First, these measures weaken a commission's capacity for self-diagnosis and self-improvement. The reason for this is that commissioners are denied the opportunity to meet informally with one another to discuss off the record the goals they want to achieve and the means by which they intend to achieve them. The realities of OMAs are that commission meetings tend to be highly structured, and they consist of a seemingly unrelated series of independent dockets. Such an arena does not facilitate unstructured discussions whereby the philosophies of individual regulators can emerge quickly and clearly. Since commissioners have only a limited ability to meet privately, such discussions seldom occur and such philosophies are not revealed efficiently.

Second, commission decisions should be both correct and legitimate. The standards of legitimacy established by APAs and OMAs are difficult to meet, but failures to meet them are fairly easy to discover. The standards of correctness are also very difficult to meet, but failures to meet them are fairly difficult to discover. Commissions, therefore, find themselves able to survive while performing poorly on the correct/incorrect scale, but a single significant lapse below the highest standards of legitimacy seriously threatens the organization. In short, regulatory agencies can survive many bad decisions, but they cannot risk even a single illegitimate one.

Those who see APAs and OMAs as important aids in making good decisions rather than as procedures for legitimizing decisions, whether they be good or bad ones, will disagree with the emphasis of my remarks. A third consequence further supports my position.

Third, in our society, the principal method for legitimizing government decisions that involve conflicting interests is the adversary method. This may reflect Americans' respect for their court system or it may merely reflect the limited imaginations of lawyers who have designed and managed much of American government. Whatever the merits of this system for settling disputes between two parties, or among a small number of parties, or in settling disputes between private parties and the government—and I am willing to grant the merits are great—it is seriously deficient in the regulatory context in two respects.[4] First, it is deficient when the objective is to discover information about the external environment that is not known to any party. The peculiar strength of adversarial proceedings is that from them knowledge about views, past actions, and, sometimes, the motives of a party can be extracted. What a commission frequently needs, however, is knowledge that no one has. To create that kind of knowledge, one must assemble experts in an environment where they freely exchange and test one another's views. Second, the adversarial process is also deficient as a legitimizing process when many of the interested parties cannot be present or effectively represented.

This point deserves elaboration. A decision-making body may use an adversarial process without being wholly dependent upon it for its information. Legislative bodies do so, and for decades economic regulatory bodies did so also. It was and is a valuable tool. With the rise of APAs, and especially with the post-Watergate procedural reforms such as OMAs, the adversarial process has tended, however, to become almost the only information-gathering tool considered legitimate. In so doing, it has tended to impose on economic regulators a passive, judge-like role.[5] Regulatory bodies, therefore, increasingly find themselves in the court-like posture of limiting the basis of their decision-making knowledge to the official record of the case—that is, to the testimony and cross-examination of the expert witnesses gathered in adversarial proceedings.[6]

Those who want to control a commission's decisions by controlling such a record obviously must insist that other channels of information to commissioners be restricted, and where they cannot be restricted, knowledge thereby gained must be deemed unusable. Consequently, commissioner contacts with industry personnel, with expert witnesses, including commission staff members in that role, and even with lay citizens concerned with the subject matter of the proceeding, are limited by *ex parte* rules and by the record-evidence requirement. To the extent, in fact, that commissioners may be generally informed, or even may be the experts they sometimes purport to be, they are increasingly told by the courts that they may not impart that general awareness or expertise into particular cases unless some basis for it has been put there by the lawyers. A standard bureaucratic joke is: "We're being treated like mushrooms. We're kept in the dark and covered with manure." In the case of regulatory commissions, they are kept in the sunshine and covered with legal arguments. The differences are not significant. In short, a commissioner must not merely use the record, he or she must increasingly, deliberately cultivate ignorance of off-the-record material.

The ignorance is not, however, totally self-imposed. Existing mechanisms designed to serve passive decision makers are poorly designed to bring regulatory attention to certain problems until they reach crisis proportions and/or appear in formal records of the organization. Passive decision makers are expected to wait for disputes to be brought to them; they are not expected to search the environment looking for potential disputes. The financial situation in which several electric utilities find themselves today is an example of this problem. It can be said that these utilities arrived at a form of no-fault bankruptcy with only a minimum of regulatory attention. But as a regulator, I only know that because I read the papers. I would not be aware of that if I relied on seeing it in the record of a FERC case.

With regulatory commissions tied to an official record, the battle for commissioners' minds becomes a battle for dominance of the record. In such a battle, well-organized, well-financed interests with a great deal at stake in a given case have a comparative advantage. More troublesome, and more relevant to my point, is that commissioners tied to an official record find it difficult to argue that they have the ability to make decisions in the public interest, if that interest is not defined and represented in the record. It should not surprise knowledgeable observers, therefore, to see increasing concern among certain parts of the public that their interests are not being protected by public utility commissions.

In state after state we have seen a growing battle for dominance of the record. Attorney generals intervene in commission proceedings claiming that they must represent the public interest because the commission cannot or will not. State consumer advocate offices have been created, sometimes to represent the public interest but more often to represent interests of residential customers because, it is argued, the commission will not or cannot protect those interests. For several years the federal government, through legal aid programs, financed intervention to protect interests of the poor and elderly.[7] Recent years have seen the rise of two Citizens' Utility Boards (CUBs).[8] Several more are being proposed.

In Illinois, the state with which I am most familiar, the principal argument used by CUB proponents was that the commission was bound by the record and the record was dominated by utilities and large utility customers. Consequently, residential and small commercial interests could not be adequately protected by the commission. The Illinois General Assembly, aware of public dissatisfaction with the commission and its decisions, and in order to fend off proposals to elect commissioners, approved the creation of the CUB and provided financing to help it organize.

There are related reasons for the extreme degree to which the decision-making process has been captured by the legal profession. First, it is a natural extension of the fact that commissions' most immediate critics and reviewers are judges. But it also represents the differing viewpoints of what the commissions' purposes are. Lawyers are trained to look for justice. Economists, on the other hand, are trained to look for efficiency. But the extreme legalization of the process (the FERC staff has some 200 attorneys and fewer than 50 economists), the legal method of reducing issues to the most narrow and most legalistic form, and everyone's attempt to gain professional home field advantage, all too often lead to records presented for consideration that are stripped of almost all economics or any other generalized information that would permit a decision to be made in its full context. Even when one or more parties does include testimony about the economic environment in a record, it is often lost on its bureaucratic travels toward a commission decision. It may be lost because the initial drafters of orders, emphasizing their intellectual strengths, emphasize the legal issues of the case, or it may be lost because the decision-making process is often so long that the economic environment changes noticeably between the time the record is created and the time the decision is made.

In summary, the heavy reliance on court-like adversarial proceedings, court-like *ex parte* standards, and court-like decision-making processes to legitimize regulatory commission decisions, aggravated as they are by OMAs that severely limit communication between and among commissioners, may have served to legitimize decisions in the short run, but they are slowly undermining the legitimacy of the role of commissions as protectors of the public interest.

A related problem arises in non-rate-making issues with which the FERC must sometimes deal. These cases usually require the commission to allocate public resources or license certain activities. Precise criteria and directives embodying social and political decisions are often given to the commission for implementation. Thus, the commission is often pushed into a damage control operation in which the political overlay of noneconomic criteria inevitably cause decisions that are inconsistent with any attempt to develop and implement sound economic principles. The allocation of hydro-power in New York is an outstanding example of this problem. The concentration of benefits to a small body of beneficiaries induces the maximum cupidity of others. The apparently unbounded ability of parties to sue each other to capture a piece of these economic rents promises a chain of litigation that will end only when all the economic rents have been captured by lawyers. The lawyers in these cases often tell commissions that they are compelled to obey the most precise provisions of a statute, even at the expense of its general purposes, rather than treating it as courts might treat a trust indenture, whose specifics may have to yield in order to accomplish the fundamental purpose for which it was created.

Another view of commission decision-making processes complements the preceding argument. APAs and OMAs have severely complicated the task of synthesis. The essence of synthesis is defining and evaluating alternatives. A "before and after" example can illustrate this problem. Several years ago, before Illinois adopted an APA and imposed a rigid OMA on the Commerce Commission, the task of synthesis was relatively easy, that is, easy compared to what it later became. As a case was proceeding, the commissioners met with one another, with senior staff members, and with the hearing examiner as the need arose. In those meetings, alternatives were defined and evaluated. In defining and evaluating alternatives, commissioners could efficiently exploit the knowledge of those intimately familiar with the official record, such as the hearing examiner and occasionally staff witnesses in the case. It also had in those meetings senior staff members who could provide an historical context and related the current decision to other past and developing decisions. These experienced staff members and the experience of long-tenured commissioners clearly identified the policy implications of the decision. This process also facilitated a division of labor among commissioners.

While only five commissioners voted and always cast the final votes in a public session, the decision was truly a team decision by about ten people. The five or so senior staff members might vary from case to case.

After the implementation of the Illinois APA and OMA, the decision-making process became quite different. The OMA made closed meetings of a "majority of a quorum" of commissioners difficult. Consequently, there were few. The division of labor among commissioners was substantially reduced as each commissioner became a much more free-standing entity. Collective wisdom, which I had learned to respect, was largely destroyed. The APA greatly limited communications between commissioners and commission staff. Official records became longer: 14,000–15,000 pages in a few cases, 3,000–5,000 in many. Commissioners acting individually with the help of their small personal staffs and the hearing examiner now had to carry most of the burden of defining and evaluating alternatives. Senior staff members had to try to educate commissioners one at a time. All too often the result was a single proposal presented to the commission, and this proposal was often insufficiently evaluated. Decision-making power tended to shift to the staff and hearing examiners.

The APA and OMA greatly expanded the importance of the adversarial parts of the decision-making process and the official record. In doing so, they substantially increased both the quality and quantity of analysis in the record. At the same time that analyses of particular parts of problems were improving, the ability of commissioners to use the expertise of the agency to synthesize that knowledge and put it in a useful context was severely damaged by the new communications barriers. The Illinois commission was still struggling with that problem when I left it in July 1984. I expect that it will be for quite some time.

One of the lessons I have learned at the FERC during my short tenure is that the contrast between sophistication in analysis and deficiencies of synthesis is even more extreme in the FERC than it was in the Illinois commission. The FERC is an agency with a truly impressive array of staff talent. Consequently, the quality of analysis on many topics is also impressive. This array of talent means that the division of labor within the staff is much more elaborate than I was accustomed to in Illinois. It is a textbook principle that the more elaborate the division of labor in an organization, the greater is the interdependence of the parts and the greater is the need for coordination and communication.[9] Any deficiency in this coordination and communication makes the task of synthesis unnecessarily difficult for the key decision makers.

My current view is that the task of synthesis is not being done well at the FERC. Perhaps this view reflects only my inexperience and the frustration of trying to gain a grip on the massive case load, but I doubt it is only that. The Federal Sunshine Act encourages free-standing commissioners, just as the OMA did in Illinois. The federal APA severely limits commissioners' (and their assistants') access to key staff members on crucial matters. In certain ways this restraint is even more binding than it was in Illinois. And the official records are enormous, inhumanly enormous.

The very size of the organization and the elaborate division of labor within it make it imperative that the task of synthesis be given highest priority, but traditionally such work is left largely to commissioners and their assistants. Several problems flow from this lack of synthesis.

- While the FERC is sufficiently well organized to provide each commissioner with detailed knowledge of each case before the commission, it is not organized to provide useful and timely information about the industry and the economic context for the case.
- The elaborate division of labor in the commission, the large volume of cases, and the tedious and time-consuming nature of the hearings process has induced the commission to segment issues for administrative reasons that are essentially indivisible—for example, to determine rate structure in one proceeding and rate of return in another. Regulated firms often initiate actions of this sort as well. Consequently, the commission frequently finds itself being required to make sequentially those decisions that logic and good economics say should be made simultaneously.
- The commission finds it difficult to develop a consensus about the nature and direction of evolution of the industries it is charged with regulating. The forest and trees problem is all-pervasive.
- Even when the synthesis does occur, the decision-making structure demands heroic measures to implement a consistent sequence of actions. Change, therefore, either innovative or adaptational, comes remarkably slowly. When I ask why the staff recommends a certain action, I am frequently told that the action is consistent with what the commission decided in past years, and staff will keep on doing it that way until the commission tells it to change. An organization whose central purpose for many years was to do cost-based regulation on a monopoly utility with a relatively inelastic demand has now been asked to apply its regulation to industries that are increasingly competitive.

Because this problem of synthesis is so difficult and so poorly mastered by the FERC, it is difficult to imagine consistently constructive leadership by the FERC on the application of good economic principles to regulated industries. This problem would be difficult under the best of circumstances, but APA rules and Sunshine Act rules aggravate the difficulty.

There are a lot of implications to be drawn from this conclusion, but I will only mention one. Both the gas and electric industries are becoming more competitive. The importance of distinguishing between constructive and nonconstructive competition, and setting rates and terms of service that encourage the former and discourage the latter cannot be overestimated. Encouraging competition with regulated rates can produce the evils of both regulation and competition and the benefits of neither unless done with substantial economic sophistication. The FERC does not now have that sophistication either in the dimension of knowledge or in the dimension of organization. Gaining that knowledge will be difficult. Gaining that administrative sophistication may be impossible within American traditions of procedural propriety.

NOTES

1. An anonymous lawyer quoted in S. Lawrence Paulson, "Producers Wonder How to Deal with FERC's New Role in Gas," *Oil Daily*, October 29, 1984.

2. The following discussion draws on my experiences as a commissioner on the Illinois Commerce Commission and on the Federal Energy Regulatory Commission (FERC). Many of the conclusions are generalized because other commissions have structures and decision-making processes similar to those on which I have served. Many years of discussing these problems with other commissioners, both state and federal, encourage me to believe that such generalizations are defensible.

3. This failure to agree upon purposes makes it difficult to gain effective voluntary cooperation from professional staff members: "The idea of individuals in an organization being self-starting and self-directing in providing coordination is appealing. But a number of things frustrate the use of this form of coordination, particularly in larger organizations. To have voluntary coordination, an individual has to have some knowledge of the goals of his unit, his position, and the conditions internal and external to the organization which have to be accommodated." See Joseph A. Litterer, *The Analysis of Organizations* (New York: John Wiley and Sons, 1965), p. 223.

4. For a brief defense of the adversary system see Stephen Landsman, *The Adversary System: A Description and Defense* (Washington, D.C.: American Enterprise Institute for Public Policy Research, 1984).

5. While economic regulators are not yet as constrained in their decision-making behavior as are judges, it is worth noting the type of behavior considered desirable in a strict adversary system.

The adversary system relies on a neutral and passive decisionmaker to adjudicate disputes after they have been aired by the adversaries in a contested proceeding. He is expected to refrain from making any judgments until the conclusion of the contest and is prohibited from becoming actively involved in the gathering of evidence or the settlement of the case. Adversary theory suggests that if the decisionmaker strays from the passive role, he runs a serious risk of prematurely committing himself to one or another version of the facts and of failing to appreciate the value of all the evidence. . . . One of the most significant implications of the adoption of the principles of nuetrality and passivity is that they tend to commit the adversary system to the objective of resolving disputes rather than searching for material truth. Ibid., p. 3.

6. Even the courts have recognized the danger of an administrative agency charged with protecting the public interest relying too heavily on a strict adversarial process. See *Scenic Hudson Preservation Conference v. FPC*, 354 F.2d 608, 620 (2d Cir. 1965): The commission may not merely "act as an umpire blandly calling balls and strikes for adversaries appearing before it; the right of the public must receive active and affirmative protection at the hands of the Commission."

7. The FERC was authorized six years ago to create an Office of Public Participation. See 16 U.S.C. 825q-1 (1982). The office was designed to assist the public's participation in commission proceedings and is authorized to pay intervenors' costs, including attorneys' and witnesses' fees. After legally requiring that "there shall be an office in the Commission," the Congress never appropriated funds for the office so it does not currently exist.

8. Wisconsin created a CUB in 1979 and Illinois created one in 1983.

9. Litterer, *Analysis of Organizations*, p. 181.

27

A Critique of
Proposed Changes

Alfred E. Kahn

My perspective on the problems of the electric utility industry is dominated, and I fear therefore in some degree distorted, by my own recent experiences with the very special, exaggerated problems of some of the companies that are threatened with being swallowed up by nuclear plants they either have abandoned or are trying to complete. The group by no means includes all of the companies completing such plants—only the basketcases, such as LILCO, Public Service of Indiana, Public Service of New Hampshire, Consumers' Power, Central Maine (on whose behalf I have been a witness before the Maine state regulatory commission), and Cincinnati Gas and Electric. I must warn you, therefore, that I may well exaggerate the generality of the problems on which I plan to concentrate: my comments will apply only in greatly diluted degree, if at all, to nuclear plant builders such as Duke Power, Commonwealth Edison, or Florida Power and Light, and they certainly apply even less to the companies that are not constructing nuclear plants.

I will return to the question of the general applicability of my comments before the end of my chapter. I am impressed, however, with the generality of the view that the process of building large central station generating plants has broken down, and that no management, of any company, in its right mind will commit itself henceforward to that kind of multiyear, multibillion dollar exposure. What are the implications for regulatory policy of such a situation, if that characterization is even partially accurate? In seeking an answer to that question, I find it useful to introduce an historical perspective and even to look beyond the confines of the electric power industry.

Until ten or fifteen years ago, the peculiarly American institution of the regulated public utility monopoly served us very well. The quality of service was generally extremely good; productivity improvement continuous; and—this is the acid test—prices over the preceding decades had fallen in real terms: charges for local telephone service and gas rose considerably less than the consumer price index (CPI), and charges for electricity and long-distance telephoning had declined in absolute terms as well. Not surprisingly, therefore, practically no one had any bad things to say about either public utilities or their regulators.

Today, in contrast, the entire institution is in turmoil and disrepute. Companies and regulators compete with one another for the label of public enemy number two, each knowing that the loser will automatically be characterized as number one. Prices have outstripped the CPI in electricity and gas for a decade, in local telephone service during the last few years. And worse lies ahead: electricity ratepayers in many jurisdictions face the prospect of 30–60 percent increases in rates—no wonder they refer to it as "rate shock"—as new plants either do come into service and the rate base, or do not and are amortized. And even so, a handful of companies face the totally unprecedented prospect of bankruptcy. Similarly, it is widely predicted—and I see no reason to doubt—that the basic monthly charge for telephone service is in the process of doubling over the next several years.

The lesson to be drawn from this abrupt change seems to me obvious: it is forces almost totally outside the control of utility companies and regulators, preponderantly, that determine how well these industries perform. The happy record of the 1950s and 1960s, it has become abundantly clear, was the product primarily of exogenous factors. The most important of these were very modest rates of inflation in the economy at large, correspondingly low interest rates, rapid technological progress—in electricity generation and transmission, in the long-distance transmission of natural gas, in telephony, and in the production of the primary energy sources, oil and gas—and finally, very satisfactory rates of growth in the United States economy generally. These, along with the declining real prices of public utility services, translated into rapid, exponential growth in demand. One did not have to be very smart in those circumstances, whether as a regulator or a manager of one of these companies, to look good.

In the mid and late 1970s, in contrast, the outside world turned sour. Two crises raised the price of oil from $3.00 to $35.00 a barrel, and the field price of natural gas inevitably followed, despite regulation; our economy fell prey to chronic and, it appeared, accelerating inflation; the consequent explosion of interest rates was particularly painful for public utilities, because of their very high ratios of capital to output, which were aggravated by the increasingly long lead times in the construction of electric utility plants and exaggerated by the rapidly accumulating regulations introduced in the interest of environmental protection and nuclear safety; technological progress in electricity generation particularly seemed to come suddenly to a dead halt, and what had looked in the preceding decades like an unlimited potential of lower and lower costs for larger and larger plants suddenly seemed to reverse itself; and the dream that nuclear power might make electricity too cheap to meter turned into a nightmare. No wonder the prices of electricity and gas soared, and—in the context also of economy-wide stagnation—companies found themselves forced to cancel large generating projects right and left, leaving large numbers of them with multibillion dollar investments in nuclear plants that appear to be a total loss. There are not enough hats in the world, or rabbits to pull out of them, to make all these problems go away.

We have, however, inherited from the past a heavy superstructure of regulatory attitudes, procedures, slogans, and shibboleths, rarely examined in the 1950s and 1960s and, indeed, unexamined by a deplorably large number of people today, that served us tolerably well—or did little damage—when external conditions were propitious, but that are an obstacle to intelligent policy today. I refer to such practices as

- straight-line depreciation and application of the market cost of capital to the book value of investment;
- the "used and useful" doctrine, which sounds eminently sensible but has the effect of forcing utility companies to press to completion plants that would better be scrapped, and would seek to rationalize flagrant violations of the fundamental bargain underlying the public utility institution;
- inducing investors to accept a return equivalent to the cost of capital in exchange for an entitlement to a reasonable shot at earning it on all prudently made investments (with prudency assessed at the time the decisions were made, not after the fact);
- the interpretation of the concepts of intergenerational equity, and the matching of benefits and costs, in ways that produce demonstrable irrationalities and inequities.

Here, then, are some of the prominent characteristics of the regulatory process as traditionally—and currently—followed, which the pressure of events has forced us to reconsider. Some we have already been forced to alter, under pressure of inflation during the past decade; others continue to obstruct rational policy making.

1. Infrequent rate cases. This practice was once the norm, but it was abandoned in the 1970s. However, we may now have entered an era—though it could turn out to be very short-lived—in which we can safely return to that past practice; I will point out presently why I think that would be very healthy.

2. Rate decisions based on historic, embedded cost, as evaluated in past test years. When I came to the New York Public Service Commission, just ten years ago, the general notion was that unless cost figures had objective reality—were written down in the books of account, or even better, on tombstones—they were mere speculations, did not constitute "evidence," and could not be used. Today, of course, cost projections and future test years have become if not commonplace, then respectable.

3. Rates set and investment planning done on a company-by-company, service-area-by-service-area basis. Like the New York Power Authority, I question whether we can continue to think of New York as a conglomeration of separate, historically fortuitously defined service territories. Looking at a map of them, they were obviously drawn by Senator Gerry, of gerrymander fame. They make absolutely no sense, and treating them as separate entities for purposes of price setting produces quixotic price differentials that have no basis in economic logic or equity.

4. Grossly magnifying those irrationalities and inequities are the public and municipal power preferences clauses that, I think it can be argued, made an important, dynamic, promotional contribution in the past but that are terribly mischievous today. I merely call attention to the prospect of wholesale recapture of hydroelectric licenses by preferred customers from privately-owned companies, in a series of cases that will be flooding the Federal Energy Regulatory Commission (FERC) and the courts in the next decade. The prospect of municipal distribution companies or co-ops exercising this preference, in effect taking away rents from people who are paying 7–10¢/kWh and forced by those prices to practice intense conservation, in order to transfer them to people who are paying 2¢/kWh and laying electric cable under their sidewalks in order to save them the trouble of shoveling, is one that would in a rational society have led to rioting in the streets. And the applicants justify their preference as constituting a yardstick!

5. Investment decisions made essentially on the initiative and responsibility of management. So far as the application of independent judgment was concerned, regulation entered the process only after the fact, under the prudency doctrine—but any consequent disallowances of costs already incurred consisted of little more than nibbling around the edges and never came remotely close to jeopardizing the financial viability of the companies.

6. Regulation that did not, until a decade or so ago, worry about cash flow. In some jurisdictions regulators still profess not to do so, continuing to utter slogans about the inequity of phantom taxes or of charging customers for facilities not yet serving them. There was until the inflation of the 1970s no need to worry about construction work in progress (CWIP) in the rate base, or normalization of tax benefits, or the provision of other kinds of current cash support for construction programs in order to maintain tolerable coverage ratios and costs of capital.

7. Finally—my list may not be complete, but I think it sufficient—companies were not only permitted, they were actually encouraged to promote the use of electricity.

I do not think I have to explain how those traditional practices, mottos, and procedures have worked out and had to be modified in the altered circumstances of the previous decade, or to list the continuing manifestations of their breakdown: the prospects of rate shock, or "money-saving" rate increases, the irrational and inefficient behavior of rates over time as costly new plants enter the rate base and have applied to them current market costs of capital, the widespread threats of bankruptcy, allowances for funds during construction (AFDCs) running at 75–80 percent of earnings, with companies having to borrow to pay their dividends, and so on.

There is no need, either, for me to attempt a systematic listing of the ways in which it seems to me these various practices and doctrines are going to have to be reformed; many of them have already been yielding in various ways to the pressure of events over the last decade. Looking to the future, in any event, all I can do is try to discern some general directions that I think reform is going to have to take.

First, it seems to me we are going to have far more direct, active, and responsible intervention by public utility commissions in the process of investment planning. The companies can no longer afford to play blind man's bluff with them, and the commissions can no longer enjoy the luxury of making their decisions only after the fact. Central Maine Power, for example, has been taking the following position with its commission: We think Seabrook One should be finished, but we can live with a decision by you either way—either finish it or do not finish it. What we cannot live with is an unwillingness on your part to tell us what you are prepared to accept and to support. I think that a challenge of this kind should have been posed long before now to a number of regulatory commissions; we should see more in the future.

Second, this means a direct involvement of the public utility commissions also in developing and approving a financial plan adequate to the financing of the investment programs they have approved, or, to the extent their decision goes the other way, a plan for amortizing the costs of abandoned projects.

And that means, third, an agreement by commissions about the course of prices over the pertinent time period—the period of completion of the investment projects and of their entry into rate base, or the period of amortization. The only pertinent criteria of that optimal time path, I suggest, are economic efficiency—what time path of prices most closely approximates the path of marginal costs; minimization of the (discounted present value of the) total costs to ratepayers; and fulfillment of the regulatory bargain, in terms of permitting companies a fair opportunity to earn the cost of capital on all prudent outlays, successful and unsuccessful, and fairly sharing the burden of unforeseen and unforeseeable catastrophes between investors and consumers. Tested against these criteria, a refusal of regulators, as a matter of principle, ever to put any CWIP in the rate base is simply irrational; and I know of no other defensible criterion according to which any such policy would be other than ridiculous—any less so than a flat refusal ever to normalize the tax benefits of, say, accelerated depreciation. It should be obvious that I regard as similarly irrational and indefensible principled opposition to "phantom taxes" or to ever imposing on present ratepayers any costs associated with plants that do not yet serve them: what guidance does that last taboo give us for dealing with circumstances in which the plants do not serve future customers either—where, at the extreme, their costs will impose fiendish burdens on them: *there* is an interesting intergenerational equity problem.

These historic shibboleths, religiously observed, produce a behavior of prices over time that is totally indefensible in economic terms—flat until the plant comes in, then up by 30–60 percent, then down thereafter, over the lifespan of the plant. Nor does such a time profile of rates make any more sense in terms of equity.

Observe this additional, interesting implication of the first three tendencies I have just prescribed—a prior agreement by regulatory commissions to an investment plan, a financing plan, and, therefore, to an agreed upon course of prices over some period of time. If commission and company agree to a time path of rates over, say, a period of five years, presumably subject to some automatic indexation on the basis not of the company's own costs but of some exogenous measure of inflation, then they will have installed a powerful system of incentive regulation. During the agreed upon interval the company itself will benefit to the extent it is efficient and innovative, and suffer to the extent it is not. This is exactly the system the British have just devised for their newly privatized telecommunications industry. British Telecom will be obliged to hold the weighted average of its rates over the next five years to the retail price index minus three points. Observe how attractively such a system conforms to the prescription presented in Chapter 25 of moving away from a cost-plus to a price-comes-first system.

Before turning to Charles Stalon's tale of woe in Chapter 26, I must attach one major qualification to everything I have said so far—a qualification relating not to its validity but to its relevance and importance. In view of the fact that nobody is building new nuclear plants; in view of the fact that the resurrection of nuclear power, if it is ever going to occur, will require a whole series of institutional changes, five to ten years down the road, that have very little to do with public utility regulation; and in view of the fact that if, as Leonard Hyman suggests in Chapter 22, we are going to see companies build either no central station generating plants, or 400 MW coal plants (which apparently they can manage reasonably well under traditional regulatory practices and policies), or gas turbines, when and as needed—then is it not possible that most of our analyses and suggestions have been focused on fighting the previous war rather than the coming one? And insofar as some frighteningly large problems of the kind I have been describing remain, may they not be the problems of only a handful of companies?

I want to raise the question, also, of whether the asserted unwillingness of a large number of electric companies to commit themselves to anything larger than 400 MW coal-fired plants is itself a sign of regulatory breakdown and bad for the country. Considering the terrible inflation that we have experienced in the capital cost of coal as well as of nuclear plants, and the uncertainties about the course of demand for power over, say, the next decade, what are likely to be the comparative costs of building only a few of these large coal or nuclear plants over the next few years, and instead installing turbines when and as the need becomes clearer than it is now? I recognize that there is likely to be a real cost to that, in terms of our then finding ourselves relying on energy-inefficient, oil-and gas-burning generation; but if, on the one hand, that cost is limited to the period of years between the time when the public utility companies and the commissions begin to wish they had started building a base load plant some 6–8 years earlier, and the time when they can complete that plant, might not that waste be no greater than the kinds of waste (of capital) we have committed in base load plant construction during the past 6–8 years?

In any event, if for a long time utility companies build no major generating stations, that should at least provide us a source of some intellectual satisfaction: we may eventually be spared the constant populist allusions to the Averch-Johnson distortion in the case of utility companies whose stock is selling at 50 percent of book value,[1] on the assumption that they can make tremendous amounts of money by diluting the hell out of their stockholders—under the theory, presumably, of losing money on every sale but making it up on volume.

Now a word or two about Stalon's eloquent complaint about the obstacles to intelligent regulatory planning posed by the over-judicialization of the administrative process and the constraints imposed by the vogue of openness in government.

The first word is that his complaints are not only eloquent but justified.

But the second is that there remains enough of the original conception of an expert independent regulatory commission—that the job of regulation is too complex to be handled solely by the legislatures and the courts, and so must be vested in an expert agency with a wide range of discretion in carrying out the general legislative mandate—at least partially to overcome these obstacles.

As Stalon himself has pointed out in the case of the FERC, the staffs of these agencies are filled with an extraordinary number of able, dedicated public servants—many of them, to be sure, wedded to traditional procedures and policies, but others eager to respond to new challenges. A commissioner who is impatient with those hoary procedures and policies is in an excellent position to enlist the staff's active cooperation in breaking new ground. I have found nothing in the open meetings statutes or the traditional requirements of rigid separation between or among staff members engaged in litigation, on the one hand, and judges on the other, that prevents regular meetings between a motivated commissioner and leaders among the staff to discuss regulatory policies, or to think through and plan future directions; moreover, in the agencies with which I have been familiar, there was no obstacle to inviting a single additional commissioner to participate in these discussions. It helped, of course, to be the chairman.

Third, it is possible to make positive use of the open meetings to promote uninhibited discussion—indeed, if sufficiently encouraged, staff members will be happy to participate in open argument—about the merits of individual cases and of general policy directions. In the process, the public itself can on the one hand be educated about the complexity of the issues involved and on the other be impressed by the conscientiousness and expertise of the commission.

Another, indirect virtue of initiatives of this kind, suitably publicized thanks to the open meetings requirements and explicated to the press, is that they provide a useful antidote to misguided regulatory "reforms" such as are described in Chapter 25—efforts to compartmentalize further the regulatory process and further convert it into an effectively judicial, adversarial proceeding. An agency that actively pursues its original purpose, follows its original design of policy planning—and is seen to do so in open meetings—should be much more difficult to hamstring in that fashion.

NOTE

1. See H. Averch and L. Johnson, "Behavior of the Firm under Regulatory Constraint," *American Economic Review* 52, no. 5 (December 1962):1052–69.

Epilogue

Electricity's Future: Sharpening the Debates

Sidney Saltzman
Richard E. Schuler

What are the unresolved issues that could make a substantial difference for a region's economy and well-being as they are interrelated with the demand for and supply of electricity? In this final chapter we do not attempt to weigh all of the issues raised in previous chapters and judge the opposing contentions of their authors; rather, we seek to identify the sources of those divergent views together with additional information that might be used to enlighten the debates surrounding these issues. We also consider policy or regulatory proposals that would allow society to deal more realistically with uncertainties about the future of the utility industry. To the extent possible, we discuss these issues from a nationwide perspective.

We believe there are four overriding issues. First, how is the demand for electricity intertwined with a region's economy, and how rapidly can those demand patterns change? Second, acknowledging that the reliability of electricity supply is essential to the operation of a modern society, how do we insulate those supplies against precipitous price changes and disruptions? Third, what is the proper combination of supply sources for the future: large central station coal or nuclear units, and/or conservation and small-scale decentralized sources? Embodied in this issue are the uncertainties and risks regarding the public health and safety and the environmental impacts of the alternative sources of electricity. Fourth, have our private and public institutions failed us, should and how might they be changed, and how can these institutions resolve electricity supply issues that frequently yield benefits to one group of citizens at the expense of another?

By not including potential financial constraints among the issues discussed in this chapter, we do not wish to minimize the overriding importance of funding considerations. Rather, it is because we sense little disagreement about the importance of financial matters that we elected to focus on more contentious issues. However, we do note one thread that was raised explicitly by Philip Kron in his description of the adverse statewide financial consequences that would arise if even one utility were allowed to go bankrupt; Ronald Forbes reinforces this theme in his description of the increased cost of borrowing faced by the New York Power Authority following New York City's near bankruptcy in 1975. The simple point is there is some evidence that the purported intermarket spillovers do exist, and they must be factored into regulatory decisions.

ELECTRICITY DEMAND, THE ECONOMY, AND THE VULNERABILITY OF FUEL SUPPLIES

Since the question of security of oil supplies and therefore of fuel price patterns forms an integral part of forecasts of electricity demand, the first two issues are reviewed together in this subsection. Although most observers would agree that the Western world will be subject to future oil crises, not everyone agrees about the nature of those crises or about their potential impacts on the U.S. economy. A future oil crisis will have a profound effect on the electric future of New York and many other states, because the supply and price of oil influence the prices of other fuels that not only compete with electricity for end-use customers but also affect the price of electricity directly where oil, gas, or coal are used as boiler fuels. Any discussion of what is likely to precipitate such a crisis, while important, is much more speculative than an analysis of the impact of a crisis on the economies of the nation and of New York State, and it is this latter question that we focus on in this chapter.

Lawrence Klein, James Schlesinger, and Douglas Bohi and Joel Darmstadter all consider various aspects of the impacts of another oil crisis or embargo on national and regional (New York) economies. Klein and Bohi and Darmstadter appear somewhat more optimistic than does Schlesinger about the economy's ability to adjust to a crisis similar to those of 1973 and 1978. Schlesinger places a higher probability on the future occurrence of an oil shortage based on his analysis of the demand and supply of oil in the United States and in the world. He concludes that although significant reductions have occurred in the demand for oil as a result of conservation and reduced economic activity, anticipated future economic growth and decreases in the supply of relatively "safe" oil will again place Western nations at the mercy of OPEC. Schlesinger sees a significant growth in U.S. oil imports—an increase from the present 4.5 million barrels per day (mbd) to 9 mbd by the mid 1990s—and an even larger proportionate increase in the use of oil to generate electricity: from 0.6 mbd to 1.6 mbd by the mid 1990s because of the high cost of completing large coal-fired and nuclear plants.

Schlesinger outlines how another oil shortage (that is, crisis) could develop in the U.S. in the coming years even without any new political disturbances in the Middle East. By extrapolating from the most recent trends in the demand for oil in the U.S. and the supply of oil worldwide, he indicates that U.S. dependence on OPEC for its oil supply is likely to increase by the early 1990s despite our decreasing dependence on imported oil during the last few years. He also anticipates that this problem will exacerbate the imbalance in the demand and supply of electricity that he predicts will also occur in the early 1990s. This analysis suggests that shortages in oil and electricity will begin to emerge at about the same time in the coming decade. On the basis of U.S. reactions to earlier energy shortages, he believes a new "crisis" will develop.

Klein and also Bohi and Darmstadter recognize the likelihood of new oil shortages developing in the U.S. in the coming years, but they argue separately that the impact of such shortages will not be as severe as the two previous ones because the U.S. economy is now more resilient to such shocks than it was previously. Klein identifies the major reasons for this resiliency as conservation, interfuel substitution, and new sources of energy. In other words, market mechanisms can be expected to ameliorate similar oil crises if there is sufficient time for adjustments to occur. Bohi and Darmstadter argue further than OPEC is more interested now in price stability than it was in 1973–74 and 1979–80, partially because the demand for OPEC oil has become increasingly sensitive to price changes. Furthermore, changes in market institutions have altered the way oil transactions occur—the previous emphasis on long-term contracts has given way to an increasing use of the spot market and the development of related futures markets for crude oil and products. Their analysis suggests that these new market forces will introduce a form of stability into a market that might otherwise be disturbed by perturbations in supply.

All four authors recognize the stabilizing influence on oil markets of the U.S. strategic petroleum reserve. By implication, at least, they also recognize that political forces in the Middle East may minimize to some extent the ability of market forces to ameliorate disturbances in the supply of oil.

What are the implications of these analyses for the future of electricity? It is useful first to review the relationship between electrical energy and the economy. In an admitted oversimplification, Schlesinger relates the demand for electric power to three basic variables—the price of oil, gross national product (GNP), and interest rates. He contends that as oil prices increase, the competitive position of all oil substitutes, especially electric power, will improve, thereby increasing demand for electricity. He also suggests that the demand for electric power will grow at at rate about 0.5 percent higher than the growth of GNP. Based on his expectation of future oil shortages with resultant increases in oil prices and the government's forecasts of continued economic growth, Schlesinger sees new growth in the demand for electricity that he believes will quickly erode existing excess capacity. Because of high interest rates and the current regulatory environment, utilities are minimizing their investments in new power plants; this situation, according to Schlesinger, will contribute to electric power shortages in the future. He sees the importation of Quebec hydro-power as a relatively reliable option to meet a portion of the future electricity demand for New York State and for other states as well.

In contemplating the future demand and supply of oil and electricity, it is useful to examine what has happened in the recent past. Although the oil shortages of 1973-74 and 1978-79 precipitated significant energy conservation efforts and interfuel substitutions, changes in the relative prices of alternative forms of energy resulted in a relative shift in demand to electricity from other energy forms. Thus, although the period of 1973-83 was one of relatively slow economic growth, the very significant conservation efforts in the U.S. resulted in about a 20 percent reduction in total energy inputs to produce a unit of GNP, while per capita energy consumption also was reduced by about 17 percent (see Klein, Table 2.2, and Alvin Weinberg, Figure 13.1). On the other hand, electricity's share of total energy use increased as a result of its relatively slower price increases during this period and also because for some uses it has no substitutes (Klein, Bohi and Darmstadter).

There appears to be a consensus among the authors who are concerned with the economy of New York State that it has reached a turning point and that at least in the next 5-10 years, the economic picture should improve and the state's economy should grow at a rate above the national average. This improved performance can be explained in part by a leveling of regional cost differences (Meiners). Bohi and Darmstadter expect these cost disadvantages to be mitigated even more as regional energy and electricity price differences are reduced. The authors of the three chapters dealing with energy and electric power demand forecasts for New York State reach similar conclusions. The SEMP forecasts (Guinn) show a decline in the average annual rate of total end-use energy consumption of 0.3 percent a year from 1982 to 1999. On the other hand, the SEO forecasts electricity consumption to increase at an average of 1.3 percent per year during this same period, even though the ratio of primary energy consumption to gross state product is forecast to decline by 2.1 percent a year.

Forecasts by the NYPP show an average annual growth rate of 1.4 percent for electricity consumption (Johnson and Adams). Although very close to the SEO forecast of 1.3 percent, there are significant differences in how these forecasts were generated. The SEO model is based on more optimistic assumptions about the future of the New York State economy and a lower ratio of electricity use to gross state product than are the NYPP forecasts. Johnson and Adams argue that this ratio of about 0.5 percent forecast by the SEO is also significantly lower than that normally experienced elsewhere in the nation. While it is clear that different combinations of the forecast of electricity intensity and of the growth rate of the regional economy will lead to very different forecasts of the growth in electricity demand, what has been raised by several authors, but by no means proven conclusively, is whether or not there exists some cause and effect linkage between electric intensity and economic growth.

Clearly more research into these issues is necessary in order to establish which set of assumptions is more realistic and will lead to the more credible forecasts. In fact, it would be useful to institutionalize a process whereby a comparison and analysis of these two and other credible forecasting models and assumptions would be carried out on a routine basis. As Charles Guinn points out, the SEMP forecasts are required inputs into a wide range of legally constituted processes for dealing with major policy decisions regarding energy in the state. Because that entire planning process was not renewed recently by the legislature, however, a formal means of resolving these differences does not presently exist (Davis). It is important that significant differences between at least the two major energy forecasting systems in the state (that is, SEMP and NYPP) should be identified and their effects on the credibility of their respective forecasts evaluated.

Recognizing the difficulty of making accurate assumptions about the future performance of the New York State economy and given the strong interactions between electricity usage and economic activity, Johnson and Adams generated a number of forecasts based on more and less optimistic assumptions than their base forecast reported above. The range of their forecasts under these alternative assumptions is a high of 2.7 percent average annual growth rate and a low of 0.8 percent average annual growth rate for electricity usage.

Timothy Mount and William Deehan are also concerned with long-range forecasting of electricity demand, although their goals differ slightly from those of Guinn and also Johnson and Adams. In their model, Mount and Deehan estimate the increase in electricity rates as a result of the cost of coal conversion and of the construction of the two new power plants in New York State (that is, Nine Mile Point Two and Shoreham). They find that the high costs of adding these two nuclear power plants into the rate base in accordance with standard regulatory procedures significantly limit the growth of electricity demand. They also argue that their model probably underestimates these impacts because (a) the higher costs are spread over the entire state in their model rather than over specific service territories (Long Island, where Shoreham is located, contains a heavy concentration of the state's economic activity), and (b) their current model does not incorporate the dampening effects of higher energy prices on economic activity as have some earlier energy/economic models of New York State. Their model also does not account for the more complex long-run impacts on electric rates if a significant portion of those plant investments are declared imprudent by the PSC and kept out of the rate base permanently. Such an action has the initial effect of holding rates down, but to the extent that the financial community were, therefore, to view subsequent investments in utilities in New York as being riskier than before, required rates of return and therefore electric prices would increase.

Despite their conclusion that higher electricity prices are likely to have a significant impact on the demand for electricity, it is interesting to note that their base forecasts for long-term electricity demand are slightly higher than those of the SEO and NYPP. On the surface such similar forecasts might be comforting to those concerned with estimating long-range electricity demand in New York State. However, because different assumptions underlie each of these models, further investigation of each of these models is warranted. In fact, because New York and other large states in the U.S. often have more than one model available to them for forecasting purposes, understanding the reasons for different forecasts has become even more important. To promote such understanding, state governments should consider supporting study groups of objective experts to analyze forecasts and models.

What none of these authors deal with are the many, at best speculative, issues that might substantially alter the patterns of electricity demand by the year 2000, although Alvin Weinberg and Frank Huband do survey emerging supply technologies. Perhaps the most immediate changes may be brought about by personal computers and modern telecommunications, not only as a result of their own modest direct needs for electricity but, more important, in the way that they may reshape the working and living patterns of many Americans. To what extent will an increasing number of professionals be able to transact most of their business from home through electronic devices? Will they even want to work at home, or will psychological and sociological needs dictate that people leave their homes to work at some centralized location? These issues could have profound effects on the shape of the urban and rural areas of the future, and on the future demand for electricity (and for other fuels used in commuting and transportation).

As another example, the advent of a long-awaited breakthrough in battery technology might bring about a rapid market penetration of non-polluting (in terms of carbon monoxide) electric cars for urban commuting. Although this transition could imply a substantial increase in the consumption of electricity, most of the batteries would be recharged overnight during off-peak hours, thus creating a substantial lag in the demand for new generation capacity. The emergence, manufacture, and potential uses of biotechnologies point to another infant industry that may have a substantial but unforeseen impact on future electric demand.

The point of these speculations is that long range electricity demand forecasts have often been wrong because of unforeseen events. We have mentioned several emerging technologies that could wreak similar havoc with current demand forecasts for the year 2000. How can we devise electricity supply planning mechanisms that are responsive to fundamental shifts in demand patterns? The supply aspects of this issue are discussed in the following section.

ALTERNATIVE SUPPLY TECHNOLOGIES: LARGE VERSUS SMALL

This controversy is cast along several different axes: utility versus nonutility supply, centralized versus decentralized supply, technocrat versus ecologist, or Weinberg's energy traditionalist versus nontraditionalist; yet a careful reading of the papers in Parts III and IV shows that the constituencies for the advocates of these various positions are not sharply defined. For example, some environmentalists and conservationists favor nuclear power as the lesser of many evils, because of the absence of pollution associated with it. The issues seem to break down in stages: first, a debate between nuclear and coal-fired central station technologies; and then, comparisons between large central station supply technologies and conservation and small, decentralized, sometimes renewable-resource-based supply alternatives.

There may be no single best technology for an entire state, let alone across the nation as a whole. It is because there are widely differing demand and fuel availability patterns, environmental concerns, and spatial patterns of settlement across and within regions that current electricity supply mixes throughout the U.S. are so diverse. And in the face of continued future uncertainties about the relative prices of fuels, maintaining a large diversity of supply technologies may be a useful hedge.

None of the authors in this volume believes that new nuclear plants will be planned or constructed in the near future, although several reach that conclusion for different regulatory, public relations, and economic reasons. The short-term debate, then, really concerns central station coal-fired units versus conservation and decentralized technologies. Many regions with substantial reserves, like New York, will do nothing, despite Schlesinger's warning about the risks of that course in light of the current long lead time required to plan, gain the regulatory approvals for, and construct new large central station coal-fired units. Alvin Weinberg, Leonard Hyman, and Alfred Kahn each suggest that utility planning strategies might change as a result of those uncertainties. Utilities may decide to wait until demand increases, and then to meet and match it with small increments of capacity that can be constructed quickly. In order to gain rapid regulatory and environmental approvals, that strategy implies oil- or gas-fired units, frequently combustion turbines. Since electricity generated by those sources is expensive, power from large coal-fired units will ultimately be substituted but only after the demand has materialized. In a predictable world, this wait-and-see approach would be economically inefficient, but under the current uncertainties, it may be the most rational procedure. Meanwhile, conservation and decentralized sources of supply, if economically viable, will have an opportunity to make further inroads. However, as Lester Stuzin and Weinberg point out, the drive to find new or substitute sources of capacity may be slowed not only because of reduced growth in demand but also because it is becoming far more economical to extend the lives of existing units than to replace them with new ones.

It is for the long run that the debates between nuclear and coal-fired central station units again become important, together with the tradeoffs that smaller decentralized technologies may offer. We believe that concerns about the public health and safety and the environment will dominate. The resolution of these debates will dictate regulatory procedures and institutional requirements that in turn will largely determine the relative economics of these alternative sources of supply. (As an example, 30–40 percent of the total capital cost of recently completed nuclear plants represents interest during construction and other costs related to planning and obtaining regulatory approval.) It may be the success or failure of emerging modified technologies—for example, inherently safe nuclear reactors and fluidized-bed coal combustion—that may resolve many of the public health and safety issues and environmental concerns. The current excess supply of capacity that exists in many regions of the country provides the luxury of time to rethink from scratch whether and how these large-scale technologies might be designed and implemented in a more benign manner than is currently possible. American society has, at this time, a marvelous opportunity to plan for the future and that opportunity should not be squandered.

Significantly, not one author in this volume, whether through design, oversight, or pragmatism, calls for a massive wave of additional federally supported research and development to accelerate the feasibility of his or her favored class of technologies. What is called for are substantial changes in current pricing schemes and policies (Kahn, Percival, Schuler, and Weinberg—each for varying, frequently conflicting reasons) and in the regulatory and institutional framework under which electricity supply and planning have customarily taken place. We believe this to be a proper focus of concern, because existing regulatory and institutional frameworks can influence not only current supply decisions but also which technologies are likely to emerge in the future.

ENVIRONMENTAL AND HEALTH EFFECTS

Too much is known about the environmental and health effects of electric power generation to be summarized comprehensively in the four chapters devoted to these topics in this volume. Nevertheless, a number of very important issues are highlighted, and it is useful to examine their implications for public policy. The questions with which Senator Stafford framed his discussion of these issues are worth recalling: What is known and what is not known about these environmental and health effects? Should public officials wait for more complete information before taking appropriate action on these problems? Who should pay the costs of reducing pollution?

Fortunately, it is possible to legislate a reduction in the amount of pollution generated in the U.S. without necessarily incurring the severe economic impacts predicted by critics of this approach. For example, sulphur dioxide emissions have declined in the U.S. since the enactment of the Clean Air Act in the 1970s (Stafford). Furthermore, New York State implemented additional regulatory controls and was able to reduce its emissions from sulphur dioxide by about 50 percent (i.e., one million tons) during a period when electric power generation grew significantly (Hovey). We should not lose sight of the fact that we can influence the outcomes of certain situations if we can mobilize our collective political will.

At the present time, the dominant environmental issue for the electric utility industry appears to be acid deposition. Its effects in the U.S. (predominately in the Northeast, although there have been reports recently of its impacts in other parts of the country), Canada, and Western Europe, which are briefly described by Senator Stafford and by Hovey, are well known. Partial consensus also seems to exist about the sources of the sulphur dioxide pollutants that are involved in the process of acid deposition. Senator Stafford indicates that electric utilities account for about 75 percent of sulphur dioxide emissions east of and adjacent to the Mississippi River, whereas industrial processes generate less than 25 percent, and cars and trucks, about 2 percent. What apparently is not known as well are the transport paths the pollutants take once they are released into the atmosphere and their rates of transformation into acid. Hovey also suggests that data that attribute acid emissions (the cause) to increased acidity (the effect) are lacking, although he indicates there is a high probability that such a relationship exists.

Individual states that are suffering significantly from acid deposition have a special incentive to help alleviate this problem, yet any single state acting on its own could effect little relief because of the magnitude of the emissions from other states, as the data in Table 16.1 indicate (Hovey). For example, New York State has reduced its emissions by more than 50 percent to less than 500,000 tons per year in 1982, but that is a small portion of the total of the sulphur dioxide emissions produced in the U.S.—much of which may find their way to New York in some form of acid deposition.

What can and should be done about the problem of acid deposition, not only in New York State but in other states that have similar problems? Judging from recent history in this country, we may expect that when the problems caused by acid deposition are recognized broadly enough, the federal government will begin to take action. It did so in the past when water and air pollution became intolerable to a large enough section of society. The drawback in this approach is, of course, that the environmental costs associated with further delays in implementing a solution will continue to increase significantly. Estimates of the magnitudes of these costs might encourage additional politicians to begin to think about developing alternative solutions to the problems caused by acid deposition.

Fuel substitution is another approach that might be considered for dealing with the problems of pollution generated by the use of fossil fuels. Although alternative sources of energy (e.g., solar, wind) for generating electricity are becoming increasingly sophisticated, their rate of growth is such that they are unlikely to make serious inroads on the total use of fossil fuels in the next few decades. Until recently, the utility industry and many government agencies expected nuclear energy to assume an increasingly important role in the generation of electricity. Many knowledgeable observers probably would agree that the performance record of nuclear power in general is better than the impression the public has about its performance. However, the existing political climate for nuclear power clearly reduces the likelihood that many new nuclear plants will be built in the coming decades. Aside from the safety factor, the cost of new nuclear plants will undoubtedly inhibit the industry's future growth. It seems clear at this time that future supply decisions will be influenced by environmental and health effects as well as by economic and technical issues.

REGULATORY AND INSTITUTIONAL REFORMS

The analyses by Alfred Kahn, Richard Schuler, and Charles Stalon in Part VII are far-reaching, ranging from discussions of economically efficient pricing by regulators to the elimination of economic regulation; from the revision of regulatory procedures on the part of state commissions and the Federal Energy Regulatory Commission to provide more flexibility, to methods of integrating the economic and environmental regulation of electricity supplies. Yet Steve Barnett's discovery in his interviews of New York State residents—namely, that they simply do not trust electric utilities or the regulatory bodies overseeing them—may undermine attempts by any of these institutions to make the fundamental changes outlined in Part VII, given the fact that legislative approval must be gained to implement many of the proposals. Are politicians likely to approve structural and procedural changes that alter the ground rules for electric supply institutions but that were proposed and/or supported by those very same institutions the public mistrusts?

There are a variety of institutional and regulatory issues that surround the large-versus-small supply technology debate, but one perspective is laid out clearly by both Robert Percival and Alvin Weinberg. Percival claims there is too little conservation because utilities do not have the proper economic incentives to promote its installation; he suggests that structural institutional biases exist against the installation of the proper amount of conservation devices. In contrast, Weinberg suggests that we may have had too much small-scale hydro, wind, and co-generation installed in the past several years. He claims that the avoided cost standard (interpreted by many state commissions as the marginal cost of the most expensive alternatives) employed under the Public Utility Regulatory Policy Act of 1978 (PURPA) results in an unfair subsidy to those decentralized sources, since large central station utility generation is compensated, on average, only at its own historic cost.[1] Differential federal tax policies for utilities and for conservation and renewable energy sources compound this problem. Ideally, we would like to see instituted a "level playing field" in terms of tax treatment, pricing—regulated utility price structures should compare to those in nonregulated markets where conservation devices are sold—and institutional treatment. If that degree of similarity were established, underlying competitive forces and real cost differences should lead to the proper supply and conservation mix, provided there are not externalities (pollution or health hazards) or public-good problems (e.g., reliability or national security). In fact the national security perspective of reducing U.S. dependence on imported oil was one reason for initiating federal tax breaks for the installation of conservation devices. However, given the fact that advocates of both large central station generation and of small decentralized units and of conservation complain about the existing pricing policies, there could be some degree of rough justice inherent in the present, admittedly imperfect, price situation.

What is more disturbing is what we will call the emerging trust/size/control conflict. Many individuals no longer trust our large institutions because they seem unresponsive; some people favor turning to solutions to the problem of electricity supply which are of more manageable scale, and over which they feel they can exercise some control. Thus the public has increasingly favored conservation and decentralized solutions to the energy problems of the 1970s. Conversely, large institutions and governmental agencies find it very difficult to estimate, rely upon, or control the actions of a multitude of individual actors; therefore, those managers who are responsible for maintaining a reliable electric supply seem to favor the solutions that they feel they can control: large central station generating units. The result has been a form of institutional gridlock where the decision-making process has ground to a halt. Individual consumers have conserved, yet they find their electric bills rising. Utility planners and regulators in their attempts to complete new large generating facilities—which are in part responsible for rising rates—have been frustrated by consumer intervention. Fortunately, these incomplete facilities have not been essential for maintaining the reliability of the electric system over the past decade, but the frustrations of all parties involved do not augur well for any improvement in the decision-making process by existing institutions when supply shortages begin to appear in the future. And given DeLuca's observation that only once in the past forty years has energy been seen as a significant problem by the public at large, it may be unrealistic to expect that when another energy crisis arises in the future, sufficient public opinion will be galvanized to deal effectively with these institutional dilemmas. Schlesinger's periods of public complacency may far outweigh the public's capacity to perceive a crisis.

The environmental, public health, and safety concerns that are associated (or perceived to be) with this large-versus-small technology debate not only compound the trust/size/control conflict by adding a further dimension of risk and uncertainty to potential outcomes, they also generate new sets of advocates and opponents to any electric supply solution. These public health and safety protagonists are divided along geographic boundaries so that if one group of individuals wins, another group is almost certain to lose. In contrast, the trust/size/control conflict is merely a dispute over the means of obtaining an objective about which nearly everyone is in agreement and from which, if the means are successful, nearly everyone will benefit.

How environmental, public health, and safety problems complicate these issues is, from one perspective, based upon differential sets of experience and information. As an example, we have had substantial long-term experience with the operation of large fossil fuel-fired generating plants and we are quite certain about the nature and amounts of effluents they distribute; the impact on the environment and on humans of some of these effluents is far less certain. Whereas the human health and safety impacts of exposure for particular periods of time to different concentrations of particulates and sulphuric acid mist are well-known, we still have an imperfect understanding of the potential magnitude of the effects of acid rain and of an atmospheric mantle of carbon dioxide. By comparison, our operational experience with nuclear generation is relatively brief; therefore public confidence in estimates of patterns of radiation emissions from plant operation and failures and from the transport and eventual long-term storage of wastes is low. Ironically, we may have a much better grasp on the human health and safety consequences of these radiation emissions, should they occur, than we do on the consequences of emissions from coal-fired plants. In the case of coal, we know what is coming out of the stacks; we just are unsure of its impact. In the case of nuclear, we know the impacts, but we lack confidence in the estimates of the likelihood of releases.

When faced with this quandary, it is small wonder that the public has turned away from large central station coal or nuclear facilities. But as Alvin Weinberg and Alan Crane each point out, we are also beginning to learn about increased public health and safety risks inherent in tight, well-insulated homes—the adverse effects of chemicals and radiation that are trapped in energy-conserving structures. There are technological solutions to these problems—for example, heat exchangers that exhaust polluted air while retaining the heat in the building—but they are expensive and once again alter the relative economics of the choices.

As noted frequently in these pages, like central station fossil fuel-fired units, co-generation has been around for many years and the effluents are well-known; the environmental debates associated with it may be over uneven environmental regulatory treatment of these sources as compared with larger utility-erected units. On balance it is not entirely clear how incorporating environmental, public health, and safety considerations tips the scales on the trust/size/control conflict. It may merely heat up the debate on all sides.

What is clear, however, is that the widespread electrification of urban America has had a profound impact on improving the environment and health of those living in urban areas. It is almost impossible to conceive of the smoke, soot, filth, and stench that would permeate any major U.S. city today if it had to rely on turn-of-the-century energy technologies. Skyscrapers are only possible with elevators; if these were steam-powered, that would introduce more pollution into the urban atmosphere. So too would steam-powered air conditioning, which is only feasible for large buildings. The evolution of electrification, whether in home, business, or industry, has been toward moving the direct energy conversion, and therefore the source of pollution, from the point of application to a more remote though concentrated location. By focusing the primary energy conversion for an ever larger variety and magnitude of applications in electric generation stations, in the case of coal-fired units, it becomes more economical to remove a larger portion of the effluents than if those combustion sources were dispersed throughout the city. Increasingly, even those generating units have been moved outside the city, closer to sources of fuel and cooling water.

Therein lies the source of the second environmental and public health and safety problem: the winners and losers are different groups of people. Not only do urban residents gain the personal and economic benefits from having the electricity available within the city, they also have the direct effluent fallout shifted to someone else's backyard when the plants are no longer located in the city. Furthermore, the transmission lines used to convey the electricity from rural plant to the city represent an aesthetic insult that nonurban residents see little reason or benefit for accepting. Thus urban electricity users continue to decide to use that source of energy without consideration of the residual adverse impact on rural residents. And rural inhabitants continue to object vehemently to the construction of new lines and plants, displaying little sympathy for the health and economic benefits those facilities bestow on urban dwellers.[2] The only offset that exists in states where utilities pay property taxes to local jurisdictions is that municipalities immediately adjacent to the new facility will receive a tax windfall.

One possible solution to the problem, of course, may be to require that all generating facilities be located in the same area where the power is to be used; however, in many instances that may be simply inefficient both in terms of economic and overall environmental impacts. What is needed is some type of mechanism whereby the winners can compensate the potential losers;[3] accordingly, new facilities would be erected only if that compensation is of a form and magnitude sufficient to cause the majority of the adversely affected populace to acquiesce. Obviously, the urban residents would have to be willing to provide that level of compensation in order to make the erection of the facility desirable. Although most state regulatory bodies around the nation are empowered to evaluate such tradeoffs when they approve or deny permission to erect new electric supply facilities, they frequently are not equipped to deliver the implied compensation to the potential losers. Thus what they can do by law is currently restricted in practice by a potential political maelstrom.

How can we resolve these twin dilemmas that stand in the way of effective future electricity supply planning—the trust/control/size conflict and the winner/loser payoff? The first might be resolved merely by the passage of time if institutions continue to behave responsibly, move in some of the directions outlined in these chapters, and are lucky enough not to be faced with more of the disruptions that confronted the industry in the 1970s. But such a happy outcome may be wishful thinking, and Schuler has advocated even more decentralized, deregulated institutions to supply generation facilities (although that does not necessarily mean building smaller plants) as a necessary step to resolve (and defuse) this issue. Ironically, those steps would have no hope of success without the resolution of the winner/loser payoff. Whether generating plants are built by large utilities or independent entrepreneurs, their economical and environmentally sound location is an issue involving a public good and ultimately must be resolved in the public arena. Development of an effective mechanism to resolve these economic-environmental tradeoffs is essential not only for electricity supply and transmission facilities but also for many private industrial plants in metropolitan areas whose operation generates wastes that must be disposed of elsewhere. Who wins and who loses is increasingly becoming more important than the margin of victory.

NOTES

1. Note that Schuler's proposal to deregulate the prices for sources of all generation should reduce this problem since the deregulated price should approximate marginal cost.

2. Note that the problem with electricity is no different than the conflicts that surround toxic and hazardous waste disposal or even garbage and trash landfills in many areas.

3. The New York State Public Service Commission and New York Power Authority recently attempted to gain wider acceptance for a new 345 kV transmission line by offering recreational/educational grants valued at 2 percent of the line's cost to nearby communities. See NYSPC, "Opinion and Order Granting Certificate of Environmental Compatibility and Public Need—Power Authority of the State of New York, Marcy-South 345 kV Transmission Facilities," Opinion 85-2, Case 70126 (Albany, NY, January 30, 1985).

Index ————————————————

About the Editors and Contributors

John M. Adams is a supervisor of load forecasting at the New York Power Pool in Schenectady, New York, with over nine years' experience in the utility industry. He has a B.S. degree in electrical engineering from Rensselaer Polytechnic Institute.

Howard J. Axelrod is a principal with Planmetrics, Inc. in Albany, New York. He has B.S. and M.S. degrees in electrical engineering from Northeastern University, an M.B.A. degree from the State University of New York, and a Ph.D. degree in managerial economics from Rensselaer Polytechnic Institute. He has analyzed electric utility costs and economics for over a decade on behalf of both the New York State Public Service Commission and the Consumer Protection Board.

Steve Barnett is a vice president with Planmetrics, Inc. in New York City where he has studied the relationship between changes in American society and people's attitudes and behavior toward energy. He has an A.B. degree in philosophy from Antioch College and M.A., and Ph.D. degrees in anthropology from the University of Chicago.

Douglas R. Bohi is a senior fellow at the Center for Energy Policy Research at Resources for the Future in Washington, D.C. Dr. Bohi is the coauthor of four books on energy, and he was formerly a professor of economics at Southern Illinois University. He holds a B.S. degree in economics from Idaho State University and a Ph.D. degree in economics from Washington State University.

Alan T. Crane is a senior associate with the Office of Technology Assessment of the United States Congress in Washington, D.C. where he recently directed major projects reevaluating the role of nuclear power, and analyzing the use of coal, conservation, and solar energy. Formerly, he was involved in nuclear system design with Bechtel Power Corp. and Gulf United Nuclear Fuels Corp. He received a B.S. degree from Haverford College and an M.S. degree in mechanical engineering from New York University.

Joel Darmstadter is director of the Center for Energy Policy Research at Resources for the Future in Washington, D.C. where for nearly twenty years he has been involved in analyses of the varying roles of energy in the U.S. economy. He has a B.A. degree in economics from George

Washington University and a M.A. degree in economics from the New School for Social Research.

William E. Davis is deputy commissioner for operations of the New York State Energy Office in Albany, New York, where for the past eight years he has been involved in the state's energy and conservation planning process. He received a B.S. degree from the United States Naval Academy and a M.S.A. degree in computer systems management from George Washington University.

William Deehan is engaged in economic planning and forecasting with the Central Vermont Public Service Corporation. Formerly he was a research associate in the Department of Agricultural Economics at Cornell University, and he served for five years on the staff of the New York State Public Service Commission. Mr. Deehan is a candidate for the Ph.D. degree in agricultural economics at Cornell University where he holds an M.S. in natural resource economics. His B.S. is in resource and environmental economics from the University of Connecticut.

Donald R. DeLuca is director of the Office of Teaching and Research, a division of the Roper Center at Yale University, the world's oldest and largest archive of survey data. There, he has taught courses in survey research methodology, and since 1976 he has been a member of an interdisciplinary research group dealing with subjects in energy and the social sciences. He holds B.A. and M.A. degrees in sociology from West Virginia University and a Ph.D. degree in development sociology from Cornell University.

John S. Dyson is former chairman of the New York Power Authority and former chairman of E.P. Dutton, Inc., publishers. His long list of previous public service in New York State includes Commissioner of Commerce and Commissioner of Agriculture. He received a B.S. degree in agricultural economics from Cornell University and a master's degree in public affairs from Princeton University.

Ronald W. Forbes is associate professor of business and public policy in both the Nelson A. Rockefeller College of Public Affairs and the School of Business at the State University of New York in Albany, where he is an expert in municipal public finance, serving on the boards of several funds. He received an A.B. degree in economics from Dartmouth College and a Ph.D. degree in finance from the State University of New York at Buffalo.

Charles R. Guinn is deputy commissioner for policy and planning in the New York State Energy Office, Albany, New York where he has been involved in the state energy planning process. Previously he served in a wide variety of other economic planning positions in New York State. He holds B.S. and M.S. degrees in civil engineering from Pennsylvania State University and Northwestern University, respectively.

Nicholas C. Johnson is director of long range electric load forecasting at Rochester Gas and Electric Corporation and is chairman of the Load Forecasting Subcommittee of the New York Power Pool. Mr. Johnson has a B.S. degree in mathematics from Union College and a M.S. degree in applied mathematics from the University of Rochester, and he has over seventeen years' experience in utility planning and operations.

Robert A. Hiney is the senior vice president of the New York Power Authority's Department of Planning and Marketing where he is responsible for system planning, rates, power contracts, and customer services. He has B.S. and M.S. degrees in civil engineering from Tufts University and the State University of New York at Buffalo, respectively.

Harry Hovey is director of the Division of Air Resources, New York State Department of Environmental Conservation where he has been responsible for administering the statewide program for air pollution control since 1976. He received a B.S. degree in civil engineering from Rensselaer Polytechnic Institute and a M.S. degree in public health from the University of Minnesota.

Frank L. Huband is currently director of electrical engineering, and formerly a senior policy analyst at the National Science Foundation, Washington, D.C., where he has carried out a wide variety of research and analyses related to the environment, energy, and resource policies of the federal government. He received a B.S. degree in engineering physics and a Ph.D. degree in electrical engineering, both from Cornell University; he also has earned a law degree from Yale University.

Leonard S. Hyman is vice president and head of the Utility Research Group at Merrill Lynch Capital Markets in New York City. He has spoken and testified nationwide on electric utility issues, and he is the author of a book on that topic. He holds a B.A. degree from New York University and a M.A. degree in economics from Cornell University.

Alfred E. Kahn is the Julius Thorne Professor of Political Economy at Cornell University and a special consultant to National Economic Research Associates, Inc., New York City. In recent years, he has held a number of high level positions in the Federal and New York State governments, including chairman of the U.S. Council on Wage and Price Stability, chairman of the U.S. Civil Aeronautics Board under the Carter Administration and chairman of the New York State Public Service Commission. Dr. Kahn was the former Dean of Cornell's College of Arts and Sciences, and he is author of *The Economics of Regulation.* He received a B.A. degree and M.A. degree in economics from New York University and a Ph.D. in economics from Yale University.

Doris A. Kelley is a vice president of Merrill Lynch where since 1976 she has been responsible for the forecasting and analysis of broad trends affecting electric utilities and in particular nuclear facilities. She received a B.A. in economics from the University of Louisville.

Lawrence R. Klein, the 1980 Nobel Laureate in economics, is the Benjamin Franklin Professor of Economics at the University of Pennsylvania. His pioneering work in developing and applying large-scale econometric models for forecasting regional and national economies has gained him world-wide acclaim, and he has had substantial experience serving on a variety of governmental panels advising on the policy implications of his work. He received his B.S. degree in economics from the University of California at Berkeley and his Ph.D. in economics from the Massachusetts Institute of Technology.

Philip C. Kron is vice president in charge of the Energy East Department of Citibank in New York City. A certified public accountant, Mr. Kron has over twenty years' experience dealing with financial and accounting issues related to electric utilities and their construction projects. He has an A.B. degree and a M.B.A. degree from Dartmouth College.

Nancy E. Meiners is a senior economist in charge of the New York State Model at Wharton Econometrics, Inc., in Philadelphia, Pennsylvannia, where she is responsible for the production of forecasts of employment, population trends, and personal income in New York and New Jersey. She holds a B.S. degree in mathematics and a M.S. degree in economics from Florida State University and a Ph.D. degree in city and regional planning from Cornell University.

Timothy Mount is a professor of agricultural and resource economics at Cornell University. He has been involved in electricity and energy demand modeling and forecasting for over a decade, and he is currently developing a utility simulation model of the New York Power Pool. Dr.

Mount has a B.S. degree from the University of London and an M.S. degree from Oregon State University in agricultural economics, and an M.A. in statistics and a Ph.D. degree in agricultural economics from the University of California at Berkeley.

Robert V. Percival is a senior attorney with the Environmental Defense Fund in Washington, D.C. where since 1981 he has specialized in energy and toxic chemical issues. He has a M.A. degree in economics from Stanford University and a law degree, also from Stanford.

Sidney Saltzman is professor of city and regional planning at Cornell University, where his research interests include regional impacts of public policy. Dr. Saltzman was involved in constructing an early comprehensive regional econometric model of New York State designed to examine energy-economic-demographic interactions. It was one of the first large scale models to be used to generate forecasts of long-term energy demand in New York State. He received a B.S. degree in mechanical engineering from Purdue University, a M.S. degree in industrial engineering from Columbia University, and a Ph.D. degree in operations research and industrial engineering from Cornell University.

James R. Schlesinger is a member of the executive board and a senior advisor at Georgetown University's Center for Strategic and International Studies and a senior advisor at Shearson-Lehman Brothers, Inc., in New York City. He formerly held several cabinet positions in previous administrations in the federal government, including Secretary of Energy, Secretary of Defense, and Director of the Central Intelligence Agency. He received A.B., and Ph.D. degrees in economics from Harvard University.

Richard E. Schuler is professor of economics and of civil and environmental engineering at Cornell University. Recently, he served as commissioner and deputy chairman of the New York Public Service Commission. He has been involved in energy and electricity issues for over twenty-five years both at Cornell and previously as an engineer with the Pennsylvannia Power and Light Company and as an energy economist with Battelle Memorial Institute. He holds a B.E. degree in electrical engineering from Yale University, a M.B.A. degree from Lehigh University, and a Ph.D. degree in economics from Brown University.

Robert T. Stafford is United States senator from Vermont and is chairman of the Senate's Committee on Environment and Public Works. A former governor of Vermont, Senator Stafford has had a long-standing involvement with environmental issues, and he was the architect of the "superfund" legislation for cleaning up toxic wastes. He has a B.S. degree from Middlebury College and a law degree from Boston University.

Charles G. Stalon is a commissioner on the Federal Energy Regulatory Commission in Washington, D.C. Previously, he was a member of the Illinois Commerce Commission and an economics professor at Southern Illinois University. He received a B.A. degree in economics from Butler University and M.S. and Ph.D. degrees in economics from Purdue University.

William N. Stasiuk, Jr. is director of the Field Operations Management Group in the New York State Department of Health. He has over fifteen years' experience working on environmental, public water supply, sanitation, radiological health, and toxic substance management within New York State. Dr. Stasiuk received a B.S. degree in civil engineering and a M.E. degree from Manhattan College, and a Ph.D. in environmental engineering from Rensselaer Polytechnic Institute.

Lester M. Stuzin is the executive deputy to the chairman of the New York State Public Service Commission. Previously, he had over fifteen years' experience on electricity supply and planning issues at the PSC. He received a B.S. degree in electrical engineering from City College of New York.

Richard C. Toole is a vice president at Merrill Lynch since 1983, concentrating on electric utility and telecommunications issues, and he has been involved in the financial community and as a security anaylst for over twenty years. His previous positions were with Irving-Trust, Bache & Company, and E.F. Hutton. Mr. Toole holds a B.A. and M.A. in economics from Notre Dame University.

Alvin M. Weinberg is director of the Institute for Energy Analysis, a division of Oak Ridge Associated Universities. He has been involved with the nuclear power industry since its inception, and he is the inventor of the pressurized water reactor, which is used widely around the world today. Dr. Weinberg joined Oak Ridge National Laboratories in 1945 and served as its director for eighteen years. He received S.B., S.M., and Ph.D. degrees in physics from the University of Chicago.